# the book of
# spells

## sasha fenton

D&S
BOOKS

D&S Books Ltd
Kerswell,
Parkham Ash, Bideford
Devon, England
EX39 5PR

e-mail us at:-
enquiries@dsbooks.fsnet.co.uk

This edition printed 2007

ISBN 13 – 978-1-903327-51-7

Book Code: DS0155. Book of Spells

Material from this book previously
appeared in The Book of Spells.

Creative Director: Sarah King
Project editor: Claire Rone
Designer: Debbie Fisher & Co

Printed in Thailand

1 3 5 7 9 10 8 6 4 2

# contents

# introduction

Spells and rituals can be very effective, because, like neuro-linguistic programming, cognitive therapy and other psychological treatments, they focus the mind on one's problems or one's desires and can sometimes give help in areas where nothing practical can be done. Spells are a form of prayer, healing and love.

One of the first things that you will notice as you work through this book is that you can't expect to wave your hands around and say a few words to make a spell work. Some spells can be performed quite simply, but many require preparation, such as a shopping trip or hunting around at home to glean ingredients and equipment. Spells require forethought, and their efficacy may depend upon you casting the spell at the right time. A spell is a serious matter and casting one should take patience, effort and planning, because this extended concentration represents a personal investment in the success of the spell.

*Spells require equipment, ingredients and thought.*

*Keep your tools safe in a special box.*

None of the spells, rituals or ideas in this book is designed to hurt either yourself or another person. Indeed, all magical practitioners agree that deliberately harming others will always bounce back threefold on the protagonist. Some cultures do have a history of destructive magic: Africans and Aboriginal witches have been known for their ability to send harm to others. Such practitioners know how to protect themselves while performing these rituals, but educated and sophisticated people consider harmful magic immoral.

Some of the ancient concoctions that I have given in this book are only for information and amusement. Where I have suggested eating or drinking anything, it is only the kind of foods that you would expect to eat under normal circumstances, and it is the element of ritual that makes them special.

*Crossed fingers are a Christian symbol.*

Many spells require lighted candles, so please take care when doing this and don't set your house or yourself on fire. Never leave naked flames unattended. Some spells require silence and others need to be practised in semi-darkness or outdoors, so take care not to put yourself or others in dangerous situations. Use the spells in this book by all means, but use a bit of common sense as well.

To those who practise witchcraft, magic is a religion, but this is not a book about the religion of witchcraft. If you wish to bring a spiritual element to your spells, you can call on the help of any god or deity of your choice. Spells have nothing to do with devil-worship or with evil. Some magic is close to superstition, but this has a basis in normal human feelings and the need to make our world feel safe.

# hidden invocations for healing

The word 'recipe' originally meant a written remedy, invocation or spell. Books of spells were often called books of recipes or books of receipts. Even as late as a couple of hundred years ago, a recipe could still be a list of ingredients for a healing concoction rather than a cookery recipe. The word 'receipt' was also used in this way. A receipt was a piece of paper that had a remedy written down on it. Naturally, there were standard remedies for common ailments, and the alchemist or doctor would have a pile of these pre-written and ready to hand out. These notes were called 'prescriptions', as these had been pre-scribed, or already written out.

In Britain, and throughout Europe, pharmacies are usually signified by a cross – often an equal-sided Maltese cross. For many years, the only recognised medical practitioners, herbalists and so forth were monks, while one of the earliest hospices, or hospitals, was run by an order of religious knights in Malta. The Maltese type of cross was once a form of invocation to the great healer, Jesus Christ.

In the United States, the symbol outside a pharmacy or medical centre is usually the letter 'R' with a little stroke through the tail, or the letters 'Rx'. This stands for receipt or recipe, but, unbeknown to many pharmacists, this symbol also stands for an invocation to Jupiter or Jove. Doctors, alchemists, pharmacists and chemists once used to pray to Jupiter to make their recipe work before preparing a remedy.

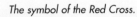

*The symbol of the Red Cross.*

Another symbol for health and healing is the caduceus, or staff, which is entwined with two snakes and topped by wings. Homeopaths and others often have this symbol on their letterhead and it is commonly seen on allergy bracelets or other health products. This is the staff of the god Mercury, who was the Roman god of healing. This symbol can still be seen engraved on a stone outside what once was the doctor's surgery and pharmacy in the ancient city of Ephesus, Turkey. This would stand as an invocation to the god Aris, who was the Greek version of the Roman god Mercury.

The Red Cross's symbol is much more modern, and while it is obviously a Christian symbol, it was actually taken from the Swiss flag. The Red Cross organisation was formed in 1861 following a war between France and Italy, aided and abetted on the Italian side by Prussia, which fixed the present-day borders between those two countries. A Swiss doctor was so horrified by the lack of medical aid for the wounded soldiers that he started the Red Cross organisation to help those who were hurt in battle. In Turkey and other Moslem countries, the same thing is symbolised by a red crescent and, in Israel, by a red star of David.

This shows that, without knowing it, the medical profession still invokes the healing powers of the deities. The cross is obviously a Christian symbol; the crescent stands for Mohammed, yet harkens back to the Egyptian goddess Isis; the caduceus stands for Mercury and the Rx for Jupiter. So when you make up, or perform, your spells, may the gods of health, healing, spells and magic always be with you!

*Candles are commonly used in casting spells, as well as in prayer (opposite).*

*The red crescent of Islam.*

*The red star of David.*

# getting started

## your magical place

You will need a place in which to work your magic that is safe, quiet and roomy enough for you to move around. Sometimes this will be outdoors, sometimes not. When you think that you have found a likely spot, sit quietly in it for a while and meditate; if it feels right, it will be right. Ensure that you are not likely to be interrupted while working your spells. Many spells require incense to be burned and candles to be lit, so check that there is enough ventilation and space for this to be done in safety. Some spells require a sacred or magic circle to be created before you can begin, and, where relevant, I have given specific instructions about how such a circle should be arranged.

If you want to use the sacred space method that today's practitioners of magic use, the following is what you should do. Find a space that

*You need a quiet place in which to work magic. To determine if you have found a good place, sit and meditate for a while.*

allows you to make a circle 9 feet (2.7 metres) in diameter. This represents the nine orders of angels and invokes the power of Mars. 'Open' your circle by walking around it in an anti-clockwise (widdershins) direction or by pointing around it with a wand, also in an anti-clockwise direction.

Now stand in the circle and imagine that you are bringing light down from the universe, so filling yourself and the space within the circle with this light. If you have a particular god, goddess, saint or other deity that you feel comfortable praying to, pray for help and guidance and for the circle to be a clean and safe place to work from. Once you have finished your work, thank the gods and ask for the light to be taken up and sent out with healing wishes to those who need help.

*Stones and crystals are tools that you are likely to need for your magic.*

Then walk around the circle, or move your wand around it, again, this time in a clockwise direction. Finally, put your tools away in a safe place and go about your normal daily business.

*Candles of different colours are useful aids for spell-casting.*

*Keep your magical tools in an attractive box.*

# tools and equipment

Many spells require special tools to make them work. These tools should be kept safely in a pretty box and should not be given to others to play around with. You could perhaps keep your container under your bed, put it in a plastic food-storage box and bury it or keep it in the garden shed, if that is a safe place. Typical tools are candles, candlesticks, incense sticks or burners, dishes, jars and bags of herbs, stones or crystals, a magic wand for making a circle, written notes – and this book. Before storing your goods, make sure that the area is clean and swept, light a candle in it and

pray for good things to happen.
You could sprinkle a little salt around the area for protection or light a stick of sandalwood incense. Meditate in the area and keep it safe and sacred for your work.

*You will need a safe place in which to keep your tools and equipment.*

# crystals

Few of us can resist buying the wonderful crystals that are now so freely available in most New Age or gift shops, but crystals need to be cleansed and energised before use. Cleanse your crystal by rinsing it in a natural source of water. If you have access to a stream of running water this is ideal, but if you don't, leave a clean bowl outside in which to catch some rainwater instead. Alternatively, you could use bottled spring water or tap water as a last resort. If you use tap water, add natural salt crystals, such as rock salt, to the water to purify it.

After cleansing, leave the crystal outside in the sun to dry off, or else on a sunny windowsill if you do not have access to a safe hotspot outside. Once the crystal is prepared, hold it in both hands and imagine white light coming down from the universe and filling your crystal with heavenly power. Ask for your crystal to bring healing, love and luck whenever it is used. If you ever feel that your crystal, or any other tool, has been invaded by bad vibes, repeat the cleansing ritual, using water and sunshine as outlined above.

You can select crystals that link with an astrological colour or the colour of a chakra. Alternatively, you can choose a crystal that has a specific energy that is attached to its colour. On the next page are some examples of crystals, along with their colours and associated energies, but these are, of course, only suggestions, and there are many others as well.

*Before using your crystal, cleanse it in water.*

# crystal energies

*Rock crystal*

**WHITE**

**moonstone, pearl, mother of pearl and rock crystal**

For cleansing the aura in preparation for working your spells.

*Rose quartz*

**PINK**

**rose quartz, kunzite**

For romantic love, self-respect, self-image; the value of anything that you are trying to achieve.

*Carnelian*

**ORANGE**

**carnelian, orange opal**

For strengthening the aura. Power, strength and courage.

*Jasper*

**RED**

**ruby, garnet, red jasper**

For energy, action, the ability to compete and sexual love.

*Citrine*

**YELLOW**

**agate, topaz, citrine, tiger's eye**

For mental activity, teaching and studying.

*Aventurine*

**GREEN**

**emerald, jade, agate, aventurine**

For healing, abundance, harmony and balance.

*Turquoise*

**TURQUOISE**

**turquoise, aquamarine**

For peace of mind, calmness and intuition.

*Lapis lazuli*

**BLUE**

**sapphire, lace agate, celestite**

For healing, protection and clairvoyance.

*Amethyst*

**PURPLE**

**amethyst, sodalite, lovulite**

In theory, these are for spiritual attainment, but I have also found them to be good healing stones.

*Tiger's eye*

**GOLD**

**gold, iron pyrites ('fool's gold'), tiger's eye, some topaz and agate stones**

For wisdom, confidence, luck and the fun side of life.

*Flint*

**BROWN**

**tiger's eye, flint, zircon**

For money, stability and abundance.

**BLACK**

**obsidian, jet, haematite**

For money, material things, protection and the ability to absorb what is good.

*Obsidian*

**AMBER**

Strictly speaking, amber is not a stone, but a fossilised resin that is hundreds of thousands of years old. This becomes obvious if you touch it with your tongue and taste it. However, it does have great healing powers, especially for chest ailments, and can also act as a calming agent.

# colour

Colour is an important factor in magic, and many spells require the use of candles or pieces of cloth, paper, string, or other odds and ends, of a particular colour. Nowadays, you can buy sets of beautifully coloured inkpens, so even if you have difficulty finding the right coloured paper for a spell, you could always use white and then write your spell in coloured ink. Colours and spells can tap into the energies of a particular astrological sign and its ruling god. Some spells are best performed on particular days of the week.

*Colour is important when weaving spells.*

# the colours and their corresponding links

### BRIGHT CRIMSON RED
**BEST DAY:** Tuesday
Any activity that requires strength, purpose, achievement, winning at anything and also sexuality.

### PINK
**BEST DAY:** Friday
Any activity dedicated to love, reconciliation, friendship or to the creation of beauty. This is also a good colour for parties, celebrations and also treating yourself and others.

### GREEN
**BEST DAY:** Friday
Green represents prosperity and abundance, but also love and partnerships of an open, rather than a secretive, nature.

### YELLOW
**BEST DAY:** Wednesday
Any activity that involves writing, phoning, e-mailing, faxing,

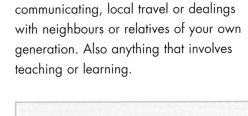

communicating, local travel or dealings with neighbours or relatives of your own generation. Also anything that involves teaching or learning.

### WHITE, SILVER, PEARL
**BEST DAY:** Monday
Any activity that concerns the domestic scene, family life, parents, property matters and everything relating to the home. This is also a good colour to use when trying to develop your intuition and the feeling side of your nature. White is excellent for dispelling evil spirits and for making a fresh start.

### GOLD, YELLOW, CREAMY COLOURS, ORANGE
**BEST DAY:** Sunday
Any activity that requires optimism and faith in the future. Gold colours concern business, creative enterprises, winning, showbusiness, music and anything relating to children. Also holidays, leisure and fun pursuits.

## BLUE
**BEST DAY:**
Wednesday
Anything
related to
health and
healing. Blue is
also useful if you

have any detailed work, exams to take or anything else that requires concentration. This is a good colour to choose for travel or dealings with foreigners.

## PURPLE
**BEST DAY:**
Tuesday
This is an
excellent
colour
for
those
who
wish to
improve their spirituality, ESP or to achieve links with higher planes. Purple also rules some very basic matters, such as joint finances, legacies, taxes, mortgages and shared resources. Also sex, birth, death and major life changes.

## BROWN
**BEST DAY:** Saturday
This colour offers protection and enhances

practicality and common sense. If you have to deal with authority figures or handle long-term enterprises or ongoing problems, choose brown. It can also link with big business, banks and money, so wear brown if you have to deal with people in these fields.

## AQUAMARINE OR TURQUOISE
**BEST DAY:** Saturday
A great colour if you are trying to float a new invention or a new and original idea. It is also useful when you want to find and make new friends or join in group activities and enhance your social life.

## BLACK
**BEST DAYS:** Tuesday or Saturday
Like white, this colour is useful for those times when you wish to bring something to an end or to make a fresh start. However, according to the ancient Egyptians, black is a very protective colour, so it might be worth wearing when you are going into a difficult situation. In addition, don't forget to ask your guardian angel for help at such important times.

# colour and the chakras

Another use of colour is the link to the 'chakras'. The word 'chakra' is Hindu for 'wheel' and refers to the wheel-shaped centres of the body that can be energised in order to get them working for magical, or other, purposes. In theory, there are said to be 78,000 chakras in the body, but there are seven main ones that are used as magical links. These are arranged in the colours of the rainbow. You may wish to enhance, or improve, the parts of your personality that link with some of th chakras and eliminate their negative aspects.

If there is some aspect of your personality, or your current thinking and behaviour, that you wish to change, focus your attention on the appropriate chakra and meditate upon it, asking your particular divinity to help you to clear the blockage, rid you of something that you are not happy with or to bring your life into balance. You can bring more power to bear on this exercise if you place an appropriate crystal on the chakra. (Don't forget to cleanse and prepare any crystal before working with it.)

**CROWN CHAKRA**
**POSITION:** crown of the head
**COLOUR:** purple
**PURPOSE:** unification of the higher self and the human personality. Spiritual will, inspiration, unity, divine wisdom, understanding, idealism and selfless service. Perception beyond space and time
**NEGATIVE:** lack of inspiration, confusion, depression, alienation, hesitation and senility

**BROW CHAKRA (THIRD EYE)**
**POSITION:** centre of the forehead, between, and slightly above, the eyebrows
**COLOUR:** indigo, dark blue

**PURPOSE:** intuition, insight, imagination, clairvoyance, concentration, peace of mind, wisdom
**NEGATIVE:** lack of concentration, fear, cynicism, detachment from reality and bad dreams

**THROAT CHAKRA**
**POSITION:** throat
**COLOUR:** sky blue
**PURPOSE:** speech and communication, creativity through speech and writing, the written word, the arts, peace, truth, knowledge, wisdom, loyalty, honesty, reliability, kindness and gentleness
**NEGATIVE:** ignorance, difficulty in

communication, depression and knowledge used for the wrong purposes

## HEART CHAKRA
POSITION: centre of the chest
COLOUR: green (in some systems, also pink)
PURPOSE: unconditional love, forgiveness, compassion, understanding, balance, acceptance, peace, harmony and contentment
NEGATIVE: repression of love, emotional instability, obsession, lack of love or harmony and lack of balance

## SOLAR-PLEXUS CHAKRA
(also known as the Pranha)
POSITION: above the navel
COLOUR: yellow
PURPOSE: will, personal power, authority, energy, mastery of desire and self-control. Warmth, humour and laughter
NEGATIVE: taking on more than one can cope with or too much emphasis on power. Hate, anger, fear and resentment

## NAVEL CHAKRA
POSITION: abdomen
COLOUR: orange
PURPOSE: giving and receiving, emotions, desire, sexual and

passionate love, change, movement, new ideas, health. Also family matters and working harmoniously with others
NEGATIVE: overindulgence in food or sex. Sexual problems. Confusion, lack of purpose, jealousy, possessiveness and impotence

## ROOT OR BASE CHAKRA
POSITION: base of the spine
COLOUR: red (in some systems, black)
PURPOSE: matters relating to the material world, success, money, mastering of the body, stability, security, courage, patience and good healt
NEGATIVE: self centredness, selfishness, insecurity, violence, greed and anger

brow chakra

throat chakra

heart chakra

solar-plexus chakra

navel chakra

root or base chakra

# elemental energies

Many magical rituals require the use of elemental energies. The elements are earth, air, fire and water. If you wish, you can add spirit to this list and thus link the elements with the five points of the pentacle. If you use the elements for spells, the following definitions may help you to understand them.

**EARTH:** strength, endurance, security, practicality
**AIR:** movement, communication
**FIRE:** transformation, energy, strong emotion
**WATER:** healing, soothing, relaxing, calming
**SPIRIT:** help from spiritual sources, increasing your ESP or healing powers

Thus, if you want to increase passion, use the fire element; if you want to achieve something practical, use earth energy; for communication, use air energy; and to calm a situation down or to give healing, use the water element.

A practical way of using these elements is to use stones or crystals for earth, incense for air, a candle for fire and a bowl of spring water for water. If you are happy to have images and idols around you, choose a picture or statuette that means something to you, such as a Buddha or a goddess, for spirit. Your image or idol can sit with your tools, and perhaps a few flowers, on a little altar.

You can absorb the element into yourself by holding it in your hands and imagining yourself being made of either fire, earth, air, water or spirit.

*The elemental altar.*

# talismans and amulets

You can also 'charge up' an amulet or talisman. To do this, hold the element in one hand and the talisman in another and then visualise the energies of the element, along with your own energies, going into the charm. This can then be given to a friend or a sick person for luck and healing. You will find further information on talismans on pages 142 to 157.

Once you have charged up your amulet, you can do things even more thoroughly. If you can cope with ancient Greek, say the following: 'Aa emptokom basum', which means 'protect me'. Below is another amulet spell, this time from ancient Egypt.

> *Great heavenly one who turns the universe,*
> *The god who is Iao, lord, ruler of all,*
> *Ablanathalaabla, grant, grant me favour.*
> *I shall have the name of the great god in this amulet,*
> *And protect me from every evil thing,*
> *Me who my mother bore and my father begot.*

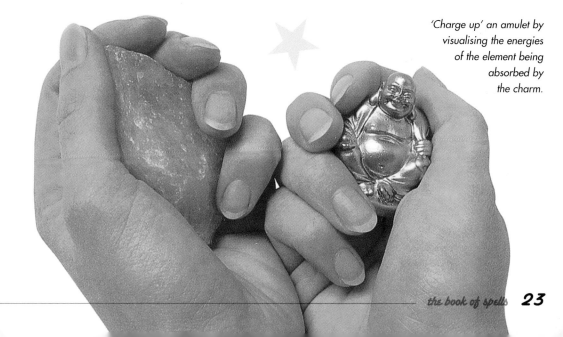

*'Charge up' an amulet by visualising the energies of the element being absorbed by the charm.*

# a time and a place...

Some spells need to be cast at specific times of the year or at particular times of the month when the moon is in the best position to ensure the success of the spell. Others are best performed on a certain day of the week or at a specific hour. The information in this chapter will help you to find the right time to ask for most things, and you can refer to it when preparing to cast spells that specify certain phases of the moon or the correct planetary day or hour for their purposes. Let us start by looking at the Celtic calendar of festivals, as these may help you to choose the right time of the year for casting some of your spells.

## celtic festivals

### 31 JANUARY
The Celtic festival of Imbolc was a lunar festival and the traditional first day of spring. This is a great time to weave spells for new beginnings.

### 21 MARCH
The spring equinox marks the time when the sun is on the equator and the days are more or less the same length as the nights. This is when the sun is increasing its power in the northern hemisphere. It was also the Celtic festival of Ostara (Easter) and it was once additionally celebrated as the ancient new year. This is a time to ask for the chance to make a fresh start or to get something important off the ground.

### 30 APRIL
The Celtic festival of Beltane links with the period just after Easter. This is a time to weave spells for relationships or projects that are progressing well, but that need an extra boost.

*Spring.*

*Summer.*

## 21 JUNE

The summer solstice and the Celtic festival of Coanhaim occur when the sun is at its strongest in the northern hemisphere and the moon is at its weakest. Solar spells for success in business, for social success or for fame should be successful at this time.

## 31 JULY

The ancient Celtic festival of Lammas, or Lughnassadh, marks the start of the harvest. The first loaf was once made from the year's wheat at this time as a mark of thanksgiving. This is a good time for spells that are designed to bring things to fruition or fulfilment.

## 21 SEPTEMBER

The autumn equinox occurs when the sun is back on the equator and the days are roughly the same length as the nights. In the northern hemisphere, this is when the moon increases in power and the sun becomes weaker. This festival was also known as Herfest, which is the probable root of the word 'harvest'. All northern-hemisphere religions celebrate a harvest festival. This is a great time of year to cast spells that are designed to bring something to the desired conclusion.

*Autumn.*

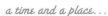

*Winter.*

## 31 OCTOBER

The Celtic festival of Samhaine is also known as Hallowe'en or All Souls' Night. On the Day of Judgement, which, according to traditional Christian belief, will occur on the following day, souls will be judged as to whether they should be resurrected or even reincarnated. Traditionally, this was the time when cattle were brought in for the winter, and those that could not be wintered were slaughtered. This is a time for letting go of the past.

## 1 NOVEMBER

All Saints' Day, as well as the future Day of Judgement when lucky souls were thought to be taken to heaven. This is a good time to ask God to preserve and keep you safe in the coming year.

## 21 DECEMBER

The winter solstice in the northern hemisphere is the time when the days are at their shortest, the sun is at its weakest and the moon is at is most powerful. This has long been considered a magical time and also a worrying one, as the ancients could never be sure that the sun was going to come back. All ancient peoples performed rituals to ensure the coming of spring. Many religious festivals involve lighting extra candles at this time of the year, and this harkens back to pagan roots. Spells for letting go of the past and for guidance about the future would work well at this time of year.

# the moon

The moon rules the rhythm of the natural world, and many spells and rituals are best performed during certain phases of the moon. Even if you are not a sky-watcher, a full moon is easy enough to spot, but you may confuse a new moon for an old one. The word 'DOC' is a useful aide-mémoire because the new moon is like the crescent of the 'D' and the old one is like the crescent of the 'C'.

For those of you who haven't the time or inclination to watch the sky, the following pages contain a lunar ephemeris (moon diary) for the next

*New moon.*

*Full moon.*

*Old moon.*

few years, up to the year 2013. This lists all of the new and full moons. I have chosen a period of seven years because seven is a particularly spiritual number and is therefore lucky for spells. In the unlikely event that you will need to find a date when the moon is a quarter full, count forwards seven days from either a new moon or a full moon. Remember that a waxing moon is one that is moving from new to full and that a waning moon is one that is moving from full to old. I have mentioned the best types of moon for associated spells at the beginning of each relevant chapter.

I have marked in the dates of eclipses, as these are especially powerful. Eclipses can have a dramatic effect on matters of love, money, career, domestic life or, indeed, anything else that is close to your heart. The Romans feared and disliked eclipses with good reason, as they can have a very nasty effect if they connect with a significant part of a person's

horoscope. An eclipse can bring a thorny situation to a head or to an end. Each eclipse has a particularly marked effect on two signs of the zodiac, which is why I have flagged these for you. However, any eclipse can illuminate a particular problem, whether it affects your particular star sign or not. Never perform spells or rituals around the time of an eclipse because the effects will be unpredictable and perhaps even the reverse of what you are trying to achieve. Full moons are said to bring extra power to spells and magic, but they, too, can have an unpredictable effect.

Some months contain two new or full moons. Such occasions are called blue moons, which is where the old saying 'once in a blue moon' comes from. There is nothing special about these as far as spells and magic are concerned, but a month containing two new moons might be beneficial.

# the days and the hours

Some of the spells in this book mention the fact that they work best if performed on a certain day of the week. This is because the days of the week are named after the seven planets that can be seen with the naked eye that were known before the invention of the

### FOR THE YEAR OF 2006

| DATE | MOON | ZODIAC SIGN |
|------|------|-------------|
| 14 Jan | Full | |
| 29 Jan | New | |
| 13 Feb | Full | |
| 28 Feb | New | |
| 14 Mar | Full eclipse | Pisces/Virgo |
| 29 Mar | New eclipse | Aries/Libra |
| 13 Apr | Full | |
| 27 Apr | New | |
| 13 May | Full | |
| 27 May | New | |
| 11 Jun | Full | |
| 25 Jun | New | |
| 11 Jul | Full | |
| 25 Jul | New | |
| 9 Aug | Full | |
| 23 Aug | New | |
| 7 Sept | Full eclipse | Virgo/Pisces |
| 22 Sept | New eclipse | Virgo/Pisces |
| 7 Oct | Full | |
| 22 Oct | New | |
| 5 Nov | Full | |
| 20 Nov | New | |
| 5 Dec | Full | |
| 20 Dec | New | |

### FOR THE YEAR OF 2007

| DATE | MOON | ZODIAC SIGN |
|------|------|-------------|
| 3 Jan | Full | |
| 19 Jan | New | |
| 2 Feb | Full | |
| 17 Feb | New | |
| 3 Mar | Full eclipse | Pisces/Virgo |
| 19 Mar | New eclipse | Pisces/Virgo |
| 2 Apr | Full | |
| 17 Apr | New | |
| 2 May | Full | Blue moon |
| 16 May | New | |
| 1 Jun | Full | Blue moon |
| 15 Jun | New | |
| 30 Jun | Full | |
| 14 Jul | New | |
| 30 Jul | Full | |
| 12 Aug | New | |
| 28 Aug | Full eclipse | Virgo/Pisces |
| 11 Sept | New eclipse | Virgo/Pisces |
| 26 Sept | Full | |
| 11 Oct | New | |
| 26 Oct | Full | |
| 9 Nov | New | |
| 24 Nov | Full | |
| 9 Dec | New | |
| 24 Dec | Full | |

telescope. The list includes the sun and the moon, which we all know are not planets at all, although for the sake of simplicity astrologers do call them planets. In Latin languages, such as French and Spanish, the planetary connection to the days of the week is obvious: for example, in France lundi is

'lunar day' or 'moon day' (Monday). English uses a mixture of Latin and Norse names, however, which makes the connections harder to spot. For instance, Monday is clearly the 'moon's day', Saturday is 'Saturn's day' and Sunday

## FOR THE YEAR OF 2008

| DATE | MOON | ZODIAC SIGN |
|---|---|---|
| 8 Jan | New | |
| 22 Jan | Full | |
| 7 Feb | New eclipse | Pisces/Virgo |
| 21 Feb | Full eclipse | Pisces/Virgo |
| 7 Mar | New | |
| 21 Mar | Full | |
| 6 Apr | New | |
| 20 Apr | Full | |
| 5 May | New | |
| 20 May | Full | |
| 3 Jun | New | |
| 18 Jun | Full | |
| 3 Jul | New | |
| 18 Jul | Full | |
| 1 Aug | New eclipse | Leo/Aquarius |
| | | Blue moon |
| 16 Aug | Full eclipse | Leo/Aquarius |
| 30 Aug | New | Blue moon |
| 15 Sept | Full | |
| 29 Sept | New | |
| 14 Oct | Full | |
| 28 Oct | New | |
| 13 Nov | Full | |
| 27 Nov | New | |
| 12 Dec | Full | |
| 27 Dec | New | |

## FOR THE YEAR OF 2009

| DATE | MOON | ZODIAC SIGN |
|---|---|---|
| 11 Jan | Full | |
| 26 Jan | New eclipse | Aquarius/Leo |
| 9 Feb | Full eclipse | Aquarius/Leo |
| 25 Feb | New | |
| 11 Mar | Full | |
| 26 Mar | New | |
| 9 Apr | Full | |
| 25 Apr | New | |
| 9 May | Full | |
| 24 May | New | |
| 7 June | Full | |
| 22 June | New | |
| 7 Jul | Full eclipse | Cancer/Capricorn |
| 22 Jul | New eclipse | Cancer/Capricorn |
| 6 Aug | Full eclipse | Leo/Aquarius |
| 20 Aug | New | |
| 4 Sept | Full | |
| 18 Sept | New | |
| 4 Oct | Full | |
| 18 Oct | New | |
| 2 Nov | Full | |
| 16 Nov | New | |
| 2 Dec | Full | |
| 16 Dec | New | |
| 31 Dec | Full eclipse | Capriorn/Cancer |
| | | Blue moon |

'the sun's day', but Tuesday, Wednesday, Thursday and Friday take their names from the Norse gods Tir, Wodin, Thor and Freya. Even so, these gods rule much the same concepts and ideas as those linked to the Roman gods.

## FOR THE YEAR OF 2011

| DATE | MOON | ZODIAC SIGN |
|------|------|-------------|
| 4 Jan | New eclipse | Capricon/Cancer |
| 19 Jan | Full | |
| 3 Feb | New | |
| 18 Feb | Full | |
| 4 Mar | New | |
| 19 Mar | Full | |
| 3 Apr | New | |
| 18 Apr | Full | |
| 3 May | New | |
| 17 May | Full | |
| 1 Jun | New eclipse | Gemini/Sagittarius |
| 15 Jun | Full eclipse | Gemini/Sagittarius |
| 1 Jul | New eclipse | Cancer/Capricorn |
| 15 Jul | Full | |
| 30 Jul | New | |
| 13 Aug | Full | |
| 29 Aug | New | |
| 12 Sept | Full | |
| 27 Sept | New | |
| 12 Oct | Full | |
| 26 Oct | New | |
| 10 Nov | Full | |
| 25 Nov | New eclipse | Sagittarius/Gemini |
| 10 Dec | Full eclipse | Sagittarius/Gemini |
| 24 Dec | New | |

## FOR THE YEAR OF 2010

| DATE | MOON | ZODIAC SIGN |
|------|------|-------------|
| 15 Jan | New eclipse | Capricorn/Cancer |
| 30 Jan | Full | |
| 14 Feb | New | |
| 28 Feb | Full | |
| 15 Mar | New | |
| 30 Mar | Full | |
| 14 Apr | New | |
| 28 Apr | Full | |
| 14 May | New | |
| 27 May | Full | |
| 12 Jun | New | |
| 26 Jun | Full eclipse | Cancer/Capricorn |
| 11 Jul | New eclipse | Cancer/Capricorn |
| 26 Jul | Full | |
| 10 Aug | New | |
| 24 Aug | Full | |
| 8 Sept | New | |
| 23 Sept | Full | |
| 7 Oct | New | |
| 23 Oct | Full | |
| 6 Nov | New | |
| 21 Nov | Full | |
| 5 Dec | New | |
| 21 Dec | Full | Sagittarius/Gemini |

## THE PLANETS AND THEIR DAYS

| | | | |
|---|---|---|---|
| The moon: | Monday | Jupiter: | Thursday |
| Mars: | Tuesday | Venus: | Friday |
| Mercury: | Wednesday | Saturn: | Saturday |
| | | The sun: | Sunday |

## FOR THE YEAR OF 2012

| DATE | MOON | ZODIAC SIGN |
|---|---|---|
| 9 Jan | Full | |
| 23 Jan | New | |
| 7 Feb | Full | |
| 21 Feb | New | |
| 8 Mar | Full | |
| 22 Mar | New | |
| 6 Apr | Full | |
| 21 Apr | New | |
| 6 May | Full | |
| 20 May | New eclipse | Gemini/Sagittarius |
| 4 Jun | Full eclipse | Gemini/Sagittarius |
| 19 Jun | New | |
| 3 Jul | Full | |
| 19 Jul | New | |
| 2 Aug | Full | |
| 17 Aug | New | |
| 31 Aug | Full | |
| 16 Sept | New | |
| 30 Sept | Full | |
| 15 Oct | New | |
| 29 Oct | Full | |
| 13 Nov | New eclipse | Scorpio/Taurus |
| 28 Nov | Full eclipse | Scorpio/Taurus |
| 13 Dec | New | |
| 28 Dec | Full | |

## FOR THE YEAR OF 2013

| DATE | MOON | ZODIAC SIGN |
|---|---|---|
| 11 Jan | New | |
| 27 Jan | Full | |
| 10 Feb | New | |
| 25 Feb | Full | |
| 11 Mar | New | |
| 27 Mar | Full | |
| 10 Apr | New | |
| 25 Apr | Full eclipse | Taurus/Scorpio |
| 10 May | New eclipse | Taurus/Scorpio |
| 25 May | Full eclipse | Gemini/Sagittarius |
| 8 Jun | New | |
| 23 Jun | Full | |
| 8 Jul | New | |
| 22 Jul | Full | |
| 6 Aug | New | |
| 21 Aug | Full | |
| 5 Sept | New | |
| 19 Sept | Full | |
| 5 Oct | New | |
| 18 Oct | Full eclipse | Libra/Aries |
| 3 Nov | New eclipse | Scorpio/Taurus |
| 17 Nov | Full | |
| 3 Dec | New | |
| 17 Dec | Full | |

# hourly planetary ephemeris

Some of the spells in this book mention the fact that they work best if they are performed on a particular day of the week or at a particular hour. The chart below shows you which planet rules each hour of the day. I have added some information about the energies of each of the planets at the end of this chapter (see pages 33 to 36), which should help you to understand why certain days and times tune into particular types of spell. Even when using spells that don't mention any specific time or day, you can choose to enhance their effects yourself by selecting a helpful hour or day.

NB: in countries such as Great Britain, where British Summer Time or Daylight Saving is in operation during part of the year, jump back one hour during the summer months.

## THE HOURLY EPHEMERIS

| HOUR | SUNDAY | MONDAY | TUESDAY | WEDNESDAY | THURSDAY | FRIDAY | SATURDAY |
|------|--------|--------|---------|-----------|----------|--------|----------|
| **AM** | | | | | | | |
| 1 | SUN | MOON | MARS | MERCURY | JUPITER | VENUS | SATURN |
| 2 | VENUS | SATURN | SUN | MOON | MARS | MERCURY | JUPITER |
| 3 | MERCURY | JUPITER | VENUS | SATURN | SUN | MOON | MARS |
| 4 | MOON | MARS | MERCURY | JUPITER | VENUS | SATURN | SUN |
| 5 | SATURN | SUN | MOON | MARS | MERCURY | JUPITER | VENUS |
| 6 | JUPITER | VENUS | SATURN | SUN | MOON | MARS | MERCURY |
| 7 | MARS | MERCURY | JUPITER | VENUS | SATURN | SUN | MOON |
| 8 | SUN | MOON | MARS | MERCURY | JUPITER | VENUS | SATURN |
| 9 | VENUS | SATURN | SUN | MOON | MARS | MERCURY | JUPITER |
| 10 | MERCURY | JUPITER | VENUS | SATURN | SUN | MOON | MARS |
| 11 | MOON | MARS | MERCURY | JUPITER | VENUS | SATURN | SUN |
| 12 | SATURN | SUN | MOON | MARS | MERCURY | JUPITER | VENUS |
| **PM** | | | | | | | |
| 13 | JUPITER | VENUS | SATURN | SUN | MOON | MARS | MERCURY |
| 14 | MARS | MERCURY | JUPITER | VENUS | SATURN | SUN | MOON |
| 15 | SUN | MOON | MARS | MERCURY | JUPITER | VENUS | SATURN |
| 16 | VENUS | SATURN | SUN | MOON | MARS | MERCURY | JUPITER |
| 17 | MERCURY | JUPITER | VENUS | SATURN | SUN | MOON | MARS |
| 18 | MOON | MARS | MERCURY | JUPITER | VENUS | SATURN | SUN |
| 19 | SATURN | SUN | MOON | MARS | MERCURY | JUPITER | VENUS |
| 20 | JUPITER | VENUS | SATURN | SUN | MOON | MARS | MERCURY |
| 21 | MARS | MERCURY | JUPITER | VENUS | SATURN | SUN | MOON |
| 22 | SUN | MOON | MARS | MERCURY | JUPITER | VENUS | SATURN |
| 23 | VENUS | SATURN | SUN | MOON | MARS | MERCURY | JUPITER |
| 24 | MERCURY | JUPITER | VENUS | SATURN | SUN | MOON | MARS |

# the sun

If you are concerned about a celebration, a special event or even the right time to set off on holiday, performing a small ritual or saying a prayer at a sun hour should bring a blessing from the sun god Apollo. A sun hour is an excellent time in which to make love, especially if you want to conceive. Spells that concern your children or young people will go well if performed at this time. The sun hour is a great time for spells that are aimed at success in business or social life, or for any event that is destined to put you at the centre of attention. The sun's association with gold and jewellery makes this a terrific time to purchase a ring or any other sparkling token of love, for yourself or your lover, or to bless such a token.

*A sun hour is an excellent itme to purchase jewellery.*

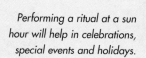

*Performing a ritual at a sun hour will help in celebrations, special events and holidays.*

# the moon

Your emotions are likely to be powerful during any moon day, moon hour or even at the time of a new or full moon, and even more so at the time of an eclipse. You may find that your intuition and ESP are heightened when the moon is making itself felt. Moon days and hours are great for performing spells that concern your home and family. Spells for finding the right home or for being able to improve or renovate one should be performed at a moon hour. Also anything that bears on family relationships should be taken on board at this time. Moon days and moon hours are great times for contacting older members of your family, or for performing spells designed to benefit them.

*The moon is for contacting older family members.*

# mars

Tackle anything that requires action and activity during a Mars hour. If you need to make a mental or physical effort at work for any purpose, get started on this while Mars rules. If you need to stand up for yourself in some situation or make yourself noticed in some way, Mars is the planet whose help you should enlist. Mars rules sex, passion and love-making, so choose a Mars hour, or at least a Mars day, for spells that are designed to bring these things your way.

*Choose a Mars hour for mental or physical activity.*

# mercury

Mercury rules communications with those who are in your immediate neighbourhood or those whom you talk to by phone, fax or e-mail. If you need to get a specific message through to someone, make a spell for this during a Mercury day or hour. Daily travel and transport are ruled by Mercury, so if your vehicle needs to be fixed, this is a good day or time to make a spell to this end. The same goes for success spells for sporting activities, although a sun day or hour are good for sports, too. Mercury is associated with healing, so if a person is suffering from ill-health, or even if your friend or loved one is unhappy and sick at heart, a Mercury day would be a good one to choose for a healing spell.

# jupiter

Jupiter was the king of the Roman gods, who could bring good or bad luck depending upon his mood. If he was in a temper, he would hurl down thunderbolts, but he also had a sunny personality and is considered to be the god of lucky breaks. Jupiter is associated with long-distance travel, so this is a great day or hour in which to make travel spells or to ask Jupiter to bless a journey for you. Business that involves overseas customers or dealing with foreigners in your own land will be successful if you enlist the aid of Jupiter. Jupiter rules justice, so a spell made to this planet is useful if you are embarking on a legal or official matter, and also before signing important documents. This planet also rules education, religion,

*Jupiter for travel, adventure or education.*

philosophy and spirituality, so a Jupiter hour can be useful for these issues as well. Some see Jupiter as a lucky planet that is associated with gambling, especially horse-racing.

## venus

Venus was the Roman goddess of love, so if you are planning a romantic meeting with a lover, or even with a potential mate, choose a Venus day or hour for your spell. If you want to propose marriage or confess love, do it during a Venus hour. Love and personal funds are closely linked in astrology, so you might wish to perform spells for prosperity or abundance during a Venus hour.

## saturn

This is a good planet to invoke if you need to bring wisdom to bear on a situation, so choose a Saturn hour if you have a sticky problem to resolve. If you need to deal with older people, father figures or those who are in positions of authority over you, performing a success spell in a Saturn hour or on a Saturn day might help.

*It is never a good idea to perform a spell during an eclipse.*

# precision is essential

The following stories show how important it is that spells be precisely worded, and they also show how the gods can have a wonderful sense of humour.

A coven worked a spell for a woman who wanted a job in a particular organisation, but instead of mentioning the organisation by name, she simply told them that she wanted a job in that particular building. She duly went for her interview, only to be turned down. Shortly after this, an employment agency sent her for another job interview, and this time she was successful. Guess what? This job turned out to be in the same building as the one she had originally gone after!

Here is a wonderful and much-documented story about a magician called Alex Saunders. Finding himself a bit short of money, he asked his coven to concentrate on getting him some gold. They duly worked the spell and a couple of days later a friend turned up at Alex's door with a gift: a bowl of goldfish.

## a motorcar spell with the wrong moon

On page 27, when dealing with the timing of spells by the moon, I mentioned that one should never perform a spell on an eclipse, and that even performing one on a full moon could bring unpredictable results. Some years ago, I visited my friend and business partner, Jonathan Dee, in his Cardiff home. At that point, he badly needed to change his car and had to find a bargain. My car was also nearing the end of its life, as were the cars of three other friends, but the most pressing problem was Jon's. Jon and I decided to cast a spell there and then, regardless of what the moon happened to be doing. As it happened, the moon was full. We went through a pentacle, paper and candle ritual out in Jon's garden at around midnight. All should have been well – and eventually it was. However, the first to get a replacement car, within less than a month, was one of our friends. Then another friend became fixed up with a car, then a third, then me . . . and it was only a couple of months after the last of those named on the list got a car that Jon got his. He has always said that it served him right for doing the spell on a full, rather than a new, moon, otherwise he would have been the first in line rather than the last!

# spells for love

The poet Lord Byron is often quoted as saying that a man may have dozens of things that he considers important, but to a woman, love is everything.

Those who consult clairvoyants, tarot card readers and so forth do so with many different problems, but matters of the heart are always high on the list. This is because, in many ways, love problems are the hardest of all to solve. Why should this be so important to clients when you consider the many real worries, like sickness, disability or major financial difficulties, that beset us? In my opinion, there is a lottery attached to matters of the heart that doesn't exist in any other area of life. Sickness is either cured or lived with, financial problems will come to some kind of conclusion, but love is such a game of winners and losers that it makes this area of life impossible to control. When one's heart is being stirred, the emotions that rise to the surface are absolutely overwhelming and can turn us into someone that even we ourselves don't

recognise. Falling in love is a kind of madness, and when an affair doesn't go the way we wish it to, then we can literally lose our heads. These runaway emotions can induce feelings of jealousy, rage, helplessness and self-hatred as a result of seeing ourselves as fools, and these emotions can astonish, and also paralyse, us to the point where we can't focus on our jobs, homes, children or friends.

A major problem is that it is impossible to make another person feel the way we want them to, or behave in the way that we would like them to, and this can lead to the kind of frustration and despair that simply does not exist in any other sphere of life. One client that all consultants hate to have to deal with is the woman who wants to get back with someone after a

parting. The chances are that her erstwhile lover was not really satisfactory and that the relationship was not well founded. Yet despite all of its faults, the client wants to turn the clock back to some place where it looked as though it would work out and where hope still existed. Spells may, or may not, help in this situation, but they have the power to calm the thoughts and to allow the person who is hurting to do something positive when no practical activity will help.

---

## CHOOSING THE RIGHT SPELL FOR LOVE

Let us now look at the differing categories of love spells so that you can choose and use those that best suit your particular situation. These may come into the categories of ending loneliness, or trying to discover who your future lover will be, or may be aimed at captivating a particular person. I have even included a spell for getting someone off your mind and out of your heart when you know that the relationship is truly over

---

## WHEN WILL I FALL IN LOVE?

It is a sad fact that for many of us, the time when we feel most ready to meet someone nice rarely coincides with it actually happening. This desire to meet someone new may arise during adolescence, after a period of study or hard work or some time after coming out of a difficult relationship, when optimism and hope start to return like the sap rising in a tree in spring. Remember that many of these spells are really ancient and therefore naturally talk of marriage. Although it would be clumsy in a book like this to change the terminology to 'relationship' or 'partnership', you can take the word 'marriage' to mean whatever you want it to.

The following ritual spells are directed to those who have no lover at present, but would like to know whether they will have one sooner, later or not at all. According to traditional belief, the first group of spells – some of which are a touch macabre – should be carried out on St Mark's Eve (24 April).

# spells for St Mark's Eve

The first spell is a mystical ceremony that is called: **'the dumb cake'**

Two or three young women (never more than three) must meet in silence to make a cake, and as soon as the clock strikes midnight, they should walk up to bed backwards, without speaking a word, for if one speaks the spell is broken. Those that are to be married will see the likenesses of their sweethearts hurrying after them, as if wishing to catch them before they get into bed. The girls should have their clothing arranged so that it can be slipped off very quickly so that they can get into bed before the apparitions catch up with them. If nothing is seen, they may hear a soft knocking at the door, or a rustling in the house, soon after getting into bed. To ensure that this is not caused by anything tangible, the girls should ensure that cats and dogs are left with neighbours.

*Dumb cake.*

# the ghostly-marriage cake

Here is another St Mark's Eve (24 April) ritual that requires the help of a friend. You must watch the church porch to make sure that there is nobody around and then place a small branch of a tree or a flower sprig inside the porch. Ensure that it is large enough to be readily found in the dark. Return home to await the approach of midnight. Just before the clock strikes twelve, and along with your friend, proceed to the porch again and remain inside it until the clock has struck. Ask your friend to pick up the flower sprig or branch and to take it to the church gate and wait there until you return. Then walk slowly towards your friend at the church gate.

If you are to be married within a year, you will see a ghostly marriage procession pass you by, with a bride who looks like you on each arm of your future husband. (Yes, this means two brides just like you!) You must also count the number of bridesmaids that appear to follow the couple, because this will be the number of months before your wedding.

# wishing cakes

A nice St Mark's Eve ritual comes from Scotland, and is easy for anyone to carry out. It involves making small cakes, but you don't need to be silent or have company while making the cakes if you don't wish to. However, you do need company when eating the cakes.
The ancient cakes that are mentioned in my old book are called 'crowdie cakes', and these may be some kind of oatcake, pancake or griddle cake. Part of the cake mixture should be set aside for those in the company who are not romantically attached, and a ring should then be placed in the mixture. Whoever gets the portion of the cake that contains the ring will be married before the year is out. I suggest that you use a gold ring for this, as you don't want to poison anybody by mistake!

Wishing cakes.

# dreaming cakes

Attracting and recognising one's future husband is clearly an important matter, and there are many spells devoted to such matters. Another version, from Scotland, is called 'bannich brauders' or 'dreaming bannocks.' Not only must you remain silent while making the cakes, but you must also add a little soot to the mixture! When the unmarried girls among the company go to sleep, they will dream of their future sweethearts. Fortunately, fresh coal soot is not poisonous, and a small quantity is actually a good cure for indigestion! One wonders whether the emphasis on silence in all these spells was just one way of keeping the girls quiet!

*Dreaming cakes.*

# salted cakes

Yet another Scots version of this ritual uses cakes called 'sauty bannocks'. These are definitely some kind of oaty griddle cake that has salt added, hence their name. This time, a charm is placed in the mixture.

# the darling buds of may

Many spells require the use of fresh plants, especially branches taken from flowering bushes; this harkens back to the idea of spring, growth and fertility.

The following is a spring spell because the idea here is to take a few branches of a flowering bush and bring them indoors. You can do this at any time after the shortest day of the year. Bash the ends of the branches with a hammer to crack them open and then put them in a vase of tepid water on a table that you will use as your altar. Now imagine a circle around you and your altar and 'charge' the area by imagining a bright-white light, with just a hint of pale green in it, coming down from high above you.

Now say the following words:

> *By stem, and by bud,*
> *By leaf, and by flower,*
> *Bring me a love*
> *In the perfect hour.*

Say this spell daily while the branches begin to bud and to flower, and the ability to love, and to attract love, will develop inside you. Believe that a lover will be sent to you, and in time (not too long, I hope) one will come your way.

# tea and sympathy

This spell requires the following ingredients:

Light the candles, put them close to your kettle and bring the spring water to the boil. Pour this into a teapot and add all of the ingredients. Leave the top off the teapot. As the tea is infusing, think of love and pass your hands over the steam three times. Pour a cup and sit in a sunny or cheerful place. If you have a photo of someone you fancy, you can place it where you can see it.

3 candles (pink, red & green)
4 caraway seeds
4 fennel seeds
1 tea bag
3 rose hips
5 edible rose petals
(that you know haven't been sprayed with chemicals)
or rose water
1 bottle of spring water

Now recite:

*I lift this cup to my lips,*
*I drink it slow*
*with tiny sips.*
*Rose, tea, caraway,*
*And fennel cause*
*love to stay.*

# hallowe'en

Long before Hallowe'en became Americanised with trick-or-treating, horror films and ghouls, it was considered a rather magical and festive time. Even now, games are popular at Hallowe'en parties, and these are probably attached in one way or another to the end of the harvest and the start of the time for storing apples and potatoes and bottling and preserving foods for the winter. One of the most popular games is 'bobbing' for apples, and while this has no specific magic connotation, it is fun. A half-barrel is filled with water and a few apples are put into it so that they float on the surface of the water. Each person is blindfolded, and he or she must put their face into the water and try to catch and lift out an apple with their teeth. This is far more difficult than it appears, because the apples keep bobbing out of the way.

The closest thing to a Hallowe'en spell that all of the young girls performed in my youth was to place a candle in a saucer on the dressing table in front of the mirror, just to one side of it. Just before midnight, we would light the candle, turn the lights off and sit looking into the mirror in the hope of seeing the face of our future husband in it.

*Hallowe'en has always been a magical time.*

*Bobbing for apples.*

*Light a candle and gaze into the mirror to see the face of your future husband.*

# love me tender

Here is a spell designed to attract a particular person. Put all of these items in your bedroom, light the candles and put them onto a table or dressing table (taking care not to set the house on fire) and lay the red scarf on your bed. Set the red ribbon aside. Place all of the other items onto the scarf, sit down quietly and allow your mind to calm down. Visualise your lover, write the following words on the heart three times and then recite them three times, addressing them to whichever god or goddess you most believe in.

> *Mix and stir and blend it so*
> *My lover's heart to wax and grow*
> *With love for me and great desire*
> *The thought of me will thrill like fire.*
> *O mote it be.*

On a Friday afternoon, preferably during the time of a waxing moon, gather together the following ingredients:

1 clover leaf
1 rose
3 red stones
A red heart (no, not a real one, just one that you've cut out of a piece of red paper or from a Valentine's card)
1 red candle
1 pink candle
A red scarf or a piece of red cotton or silk (any natural fibre will do)
13in (32.5cm) red ribbon

Gather up all of the items in the red scarf and tie the red ribbon around it, using three knots. Sleep with this under your pillow. You could also make two sets and keep one in your pocket or handbag.

*A scarf of love.*

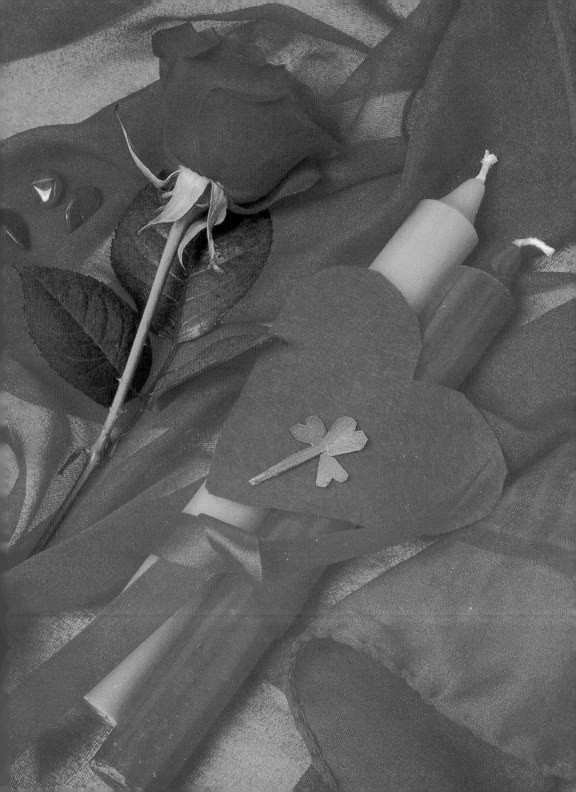

# fairies at the bottom of your garden

The idea here is to obtain the help of the fairies. This spell will take you on an outing because you will need to find a lake or a flowing stream and will have to gather your spell equipment before you set out. You will probably need to visit the library and borrow a book on trees and plants. You may also need several months of preparation for this spell, as you will have to keep back some mistletoe from Christmas.

This spell requires the following ingredients and equipment:

16 small stones or rocks (you may be able to find these at your site, but if you doubt that, take them along with you)
Plenty of flowers, a compass,
2 bowls, a rose-quartz crystal,
2 small twigs or branches,
Salt, 1 bottle of spring water,
An incense-burner,
A few dried flowers,
Leaves from different trees,
1 willow or ash wand, mistletoe

**N**

**W**

**E**

**S**

*Place your elemental objects at the compass points.*

Find your perfect spot beside a lake or a small stream. Place 16 stones in a circle that is large enough for you to sit comfortably inside. Place flowers inside the rim of the stone circle. Using your compass, locate the angles (north, south, east and west). Place the following on the angles, but just outside your stone circle.

**A rose-quartz crystal in the north.**
**A bowl of salt in the south.**
**Two twigs or branches in the east.**
**A bowl of spring water in the west.**

*Place your magical tools inside your circle of stones.*

Place your incense-burner in the centre of the circle and light it. Burn the dried flowers and some leaves. Sprinkle mistletoe on the ground and sing the following:

**I am a Druid,**
**I am a Witch,**
**Fairies, bring me love.**

**I have a wand,**
**I have a switch,**
**Fairies, bring me love.**

# where there's smoke there's fire

This spell should be performed every Tuesday after a new moon during the Venus hour. It is a kind of meditation or mantra spell.

Light the incense-burner and wait for the air above it to become smoky, then imagine your lover's face in the smoke and whisper the following over and over:

*The following equipment and ingredients are required:*

*An incense-burner*
*A small quantity of dried rose petals or yarrow*

> **Magic herbs, burn in fire,**
> **Bring to me my heart's desire.**

# when will I see you again ?

This is one of the many dumb-ceremony rituals that involves eating the yolk of an egg in silence and then filling the shell with salt. This ensures that the loved one will phone or visit before morning. The ancient source doesn't say whether the egg yolk should be eaten raw or cooked, but in today's world of health scares you should always cook the egg – particularly if you're pregnant.

# a fine-feathered spell

Collect bird feathers from your garden or a park. If you can't find any, you can buy dyed feathers from department stores. Then say the following:

> **Feathers of bird,**
> **Carry my word**
> **Into his heart**
> **Where it'll be heard.**

Then visualise yourself as the bird, visiting your lover's home and singing a pretty song to him. If you have the courage, you could then wrap one of the feathers in pink paper and send it to your lover.

# a pea-pod spell

Find a pea-pod that contains nine peas. Fix this over a door and see who comes in. If the man is a bachelor, you will have the man you want within a year. This is not necessarily the man who came through the door, however – it is only necessary for him to be a bachelor.

# a lemon spell

This is designed to let you know whether you will have the man you want. Take the peel from two lemons and carry them all day, one in each pocket. At night, rub the four posts of the bedstead with them (or rub the four corners of the bed). If you succeed, the man will appear in a dream and present you with a couple of lemons. If he does not, according to superstition, there is little or no hope.

# crystals

You can charge up a
piece of rose quartz or
moonstone by holding it
in your hands for a
while and asking the
universe to bless it. Then
ask your lover to keep it
by his bed. Some people
can wear crystals or
talismans, but others
may be allergic to them,
so keeping them under
a pillow can be a safer
alternative.

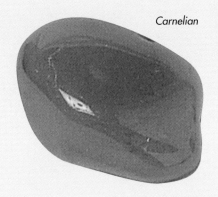

Carnelian

Amethyst

Jasper

# perfumes

Perfume has always been a powerful attractor, but not everybody likes the musk smells that are sold as agents of passion. It is said that a person who actively likes these smells is likely to be argumentative and difficult to live with.

If you want to attract or please, try a violet perfume. You could also try red or pink roses or stephanotis, which can be distilled into a perfume or used as incense.

*Persuasive perfumes.*

# flowers for fidelity

If you wish to ensure that your lover will remain faithful, why not try this idea? In the spring, either go out into the countryside and pick a few violets or buy a small pot of violets from your local florist's. Holding a few violets in your left hand, recite the following:

> **A violet breath, that opes with magic key**
> **The inmost chambers of my heart,**
> **And sets its sweetest mem'ries free.**

# washing away an obsession

When a relationship ends, it can be difficult to let go of your feelings, even if the person with whom you were involved was absolutely wrong for you. Here is a spell that will help you to banish your ex-lover from your thoughts for good.

The last item on the list at right may prove difficult, as the kind of person whom we obsess over is rarely generous enough to give presents, but any scrap of paper or any odd item that has been handed to you by them will do.

This spell requires the following components:
Fire – a candle in the correct astrological colour for the person you want to get out of your life,
Paper in the appropriate astrological colour,
Water, a little glue,
A saucer, lighter or matches,
1 bottle of spring water (optional),
A fast-flowing stream,
Something that has been given to you by the lover or something that belonged to the lover.
This must be something that can be destroyed by fire

Take the paper and cut a small strip that can be wound around the bottom of the candle. Write the person's name on the paper and a request that he or she be removed from your life and sent away to the south. (Why they should be sent to the south is a mystery, but this is how the spell proceeds!) Then wrap the strip of paper around the base of the candle and stick it with glue so that it stays firmly in place. Place the candle on the saucer near water – the ideal location is a draining board, as there is water in the taps. This spell should also be cast near a source of natural light, so if your draining board is by a window, so much the better. If you cannot use a draining board, fill a small bowl with spring water and place it near the candle arrangement close to a window. Place the object that has been given to you on the saucer, too.

Light the candle and allow it to burn right down. When it gets close to the end, unwrap the paper from the candle (without allowing the flame to go out) and then burn it, along with the scraps of given stuff and as much of the rest of the candle as possible. Finally, take the whole lot, put it into a small bag and keep it until you can get to a stream. Then throw everything into the water and watch it flow away.

*Throw your obsession into running water.*

# spicing up your love life

Some traditions suggest that you should add borage to a salad. St John's Wort is sold as a natural tranquilliser in health stores, but can also be given as a love potion. If you can find this herb in the countryside, look for it around midsummer's day and try sleeping with the flowers and leaves under your mattress, as this is supposed to enable you to dream of the man you will marry. The marriage would then take place before St Mark's Eve (24 April).

Yarrow and hemp can also be placed under a pillow for the same effect. Yarrow has always been considered to have magical properties, and it is the stalks of the yarrow plant that are used for reading the ancient Chinese oracle, the I Ching. Yarrow leaves can be held over the eyes to give second sight. Garlic is considered to be an aphrodisiac (as long as both parties eat it, of course). Garlic pills may do the trick, and are also useful against coronary artery disease and infections, due to their mild antibiotic properties. Ginseng is another obvious aphrodisiac, as is vitamin E. Spices, such as nutmeg, ginger and caraway seeds, are said to heat the blood and improve the libido. In fact, all spices are said to be useful, and even strong coffee can

have an aphrodisiac effect, if only by keeping your lover awake! Where all strange herbs and leaves are concerned, it may be preferable to keep them in your pocket or under a pillow, as that will give the desired effect without causing allergies or other unwanted reactions.

*Garlic, ginger and caraway seeds are aphrodisiacs.*

# roses for romance

Here is a rather nice form of aphrodisiac that I found in an old text.

> ***Take three roses – white, pink and red.***
> ***Wear them next to your heart for three days.***
> ***Steep them in wine for three days and then give the drink to your lover.***
> ***After he has drunk the mixture, he will be yours forever.***

Once again, in these days of modern pesticides, perhaps it is more advisable to give your lover the wine to drink and to put the rose petals under his pillow.

# spells for careers and business

> 'There is a tide in the affairs of man, that if taken at the flow will lead on to fortune.'
> William Shakespeare

## interviews and promotion

If you have a worrying interview coming up, gather together a yellow candle and a small piece of yellow paper. Write down exactly how you would like the interview to go and wrap the piece of paper around the bottom end of the candle. Light the candle and allow it to burn down, remembering not to leave it unattended. You can also make a magic circle and sit inside it, visualising the interview going the way you would like it to.

On the day of the interview, wear some blue and yellow and carry tiny pieces of yellow crystal and blue lapis lazuli, or any blue-coloured stone, in your pocket. Finally, ask your guardian angel, saint or genie to look after you.

You can use the same techniques when you want a raise or a promotion. Yellow links with Mercury, the Roman god of communication, and blue with Jupiter, who brings success and luck. Another useful god to invoke is Apollo, who brings success in enterprises, so if you want to tap into this power source, you will need a gold candle and gold paper (or white paper and gold ink). Relevant stones might be tiger's eye, amber, carnelian, or any yellow or orange crystal or, indeed, gold jewellery.

# starting a business

The best day of the week on which to start a new business or to begin a new transaction is the same as the day of the week on which you were born. However, in the case of a tricky transaction, this is not the best day of the week in which to complete it. Otherwise, Fridays and Tuesdays are best for women, and Sundays and Mondays for men.

# a spell for your career or business

In a way, this is less a spell than an incantation or meditation, but if you light a yellow candle (for commerce) and then say the following, it can't hurt:

> *He that by the plough would thrive,*
> *Himself must either hold or drive,*
> *For age and want, save while you may,*
> *No morning's sun lasts a whole day,*
> *Get what you can, and what you get, hold,*
> *'Tis a stone that will turn all your lead into gold,*
> *Therefore be ruled by me, I pray,*
> *Save something for a rainy day.*

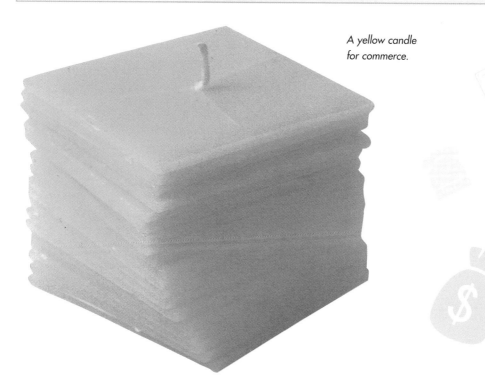

*A yellow candle for commerce.*

# another career booster

On a Thursday morning, light the green candle. Close your eyes and ask your guardian angel, saint or genie for safety and protection.

Now write your name on the paper and note down what you hope to achieve in your career. Extinguish the candle. For the rest of the day, keep the acorns, coin and paper in your pocket or handbag.

*For this spell you will need the following:*

*1 green candle*
*A piece of green paper*
*2 acorns*
*A silver coin*

On returning home from work, light the candle once again and pass the acorns and silver coin quickly through the flame.

Wrap the acorns and silver coin in the paper and bury them in your garden or in another pretty place. Repeat this spell once a year.

*A green candle to boost your career.*

# wise words

Here are some wise words on the subjects of money and business from the 18th century. Believe it or not, these old saws were collected by Napoleon Bonaparte! Strictly speaking, these are not spells, but if you light a yellow candle and write a message on a piece of brown paper, or in brown ink on white paper, and then read the message out loud to yourself, they may act as spells.

> *Remember that time is money, for he that can earn ten shillings a day at his labour and goes abroad [goes out] or sits idle at home one half of that day, though he spend but sixpence during his diversion or idleness, he ought not to reckon that the only expense.*
> *He hath really wasted or thrown away five shillings besides.*

As an author, I can only agree with this level of self-discipline and self-motivation!

> *Remember that credit is money. If a man lets money lie in my hands after it is due, because he has a good opinion of my credit, he gives interest on so much as I can make of the money during that time. This amounts to a very considerable sum, where a man hath large credit and also makes good use of it.*

Basically, this is the old adage that so many businesses follow, which is of making suppliers wait for payment.

# more wise words

> **Remember that money is a prolific or multiplying nature. Money will produce money and its offspring will produce more, and so five shillings turned is six, being turned again is seven and three-pence, and so on until it becomes a hundred pounds. And the more there is of it, the more it will produce on every turning, so that the profits rise quicker and quicker. He who throws away a crown destroys all that it might have produced, even some scores of pounds.**

My husband was in banking for over thirty years and would agree with every word of this advice, although he is not miserly enough to go along with it himself!

*Brown paper and a yellow candle will reinforce these messages.*

# other business advice

Make a note of the date, time and place that marks the start of your business. If you have a limited company, there will be articles of association, or some other official paperwork, relating to this. If you don't know what time you received these, use noon on the day that they were signed. If you have a private company, look in your book-keeping records for the first transaction that brought in money and either use noon on that day or the time that the post arrived with the cheque in it.

*Use the feng shui magic square (opposite) to bring good fortune to your home or business.*

Take this information to an astrologer, along with a note of your own time, date and place of birth, and he or she will be able to tell you what the prognosis will be and what you can do to improve matters.

**south-east**       **south**       **south-west**

| 4 | 9 | 2 |
|:---:|:---:|:---:|
| **3** | **5** | **7** |
| **8** | **1** | **6** |

**east** (left of middle row)       **west** (right of middle row)

**north-east**       **north**       **north-west**

## career area

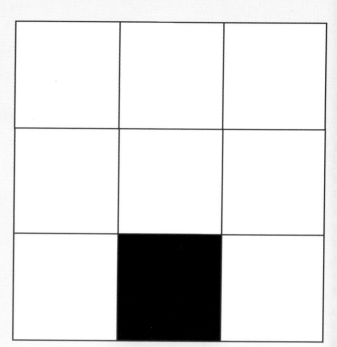

## wealth area

Alternatively, or additionally, you could treat yourself to one or two of the many available books on feng shui and prepare your workplace Chinese-style. Use the magic-square system that is in most books on feng shui and place a plant that has rounded leaves in the money and business corners of the premises. Also place some Chinese coins, a lucky Chinese toad ornament and some flowers there. If communication is an important part of your work, you could place a water feature or a tank of fish in the area where this takes place.

*A lucky toad and coins.*

# healing magic

If you or anyone else are unwell, common sense is required. You can't cure a serious disease or a broken leg by spiritual healing or magic alone, but anything that sends good vibes increases the body's ability to heal itself, so it is a good aid to standard medical treatment. Even doctors now say that alternative, spiritual and magical healing are an excellent aid to recovery from illness or operations. Doctors also agree that healing helps in the case of chronic ailments that normal medicine can't do much for. Often just the thought that someone cares enough to do something to help is beneficial. So read the following and use the advice given sensibly.

*Use an oil-burner to assist your healing spell.*

## operations and other procedures

If you are due to go into hospital for an operation or any other type of invasive procedure, where possible, choose a date during a waning moon. One would think that a waxing moon would be the best time for an operation, as this implies growth, but this influence is thought to make the blood flow too freely, therefore making the body less able to heal itself and deal with shock.

# health spells from around the world

This one comes from Finland and requires some aromatic oil to be burned on an aromatherapy burner, which are widely available. Place some cold water in the top part of the burner and add a few drops of a favourite aromatherapy oil to it. Light the tea light beneath the oil and water and recite the following:

> *Sickness, vanish into the sky!*
> *Pain, fly up into the air.*
> *Burning vapour, rise into the air!*
> *So that the wind may take thee away.*
> *So that the storm may chase thee to far places*
> *Where neither sun nor moon gives light,*
> *Where no hot wind burns the flesh.*

This is a Cherokee spell to cure a mobility-impaired person:

> *You, O Red Woman, you have caused it.*
> *You have put the intruder under him.*
> *Ha! Now you have come from the Sun Land.*
> *You have brought the small red seats with your feet resting upon them.*
> *Ha! Now they have swiftly moved away from you.*
> *Relief is accomplished.*

# healing yourself

This ritual is best performed when the moon is waning so much that it is almost becoming new again. You can either do this for yourself or you can work with a sick person, but they must

| BODY AREA | ELEMENT | |
|-----------|---------|---|
| ABDOMEN | FIRE | |
| ARMS | EARTH | |
| BLOOD | AIR | |
| BONES | EARTH | |
| CHEST | AIR | |
| EARS | WATER | |
| EYES | WATER | |
| FEET | EARTH | |
| GENITALS | FIRE | |
| HANDS | EARTH | |

do the ritual for themselves under your guidance because you can't do it for them.
First you (or the person who is sick) must choose an element that is appropriate to the part of the body that requires attention, as set out in the table below. If there is more than one afflicted area, then work on each part one at a time.

| BODY AREA | ELEMENT | |
|-----------|---------|---|
| HEART | FIRE | |
| LEGS | EARTH | |
| LUNGS | AIR | |
| MIND | FIRE | |
| NOSE | AIR | |
| SKIN | EARTH | |
| STOMACH | WATER | |
| TEETH | EARTH | |
| THROAT | AIR | |
| UTERUS | WATER | |

Prepare a special place by gathering objects representing the elements around you.

---

**EARTH:** place any crystals that you like around the area
and place a dish that contains earth or salt nearby.
**FIRE:** candles are best for this.
**AIR:** light either an incense stick or an incense- burner.
**WATER:** a dish of spring water.

---

Take a bath and dress yourself in white. Ensure that your clothing is loose and comfortable. Draw an imaginary circle around you and place the elemental objects around the circle. If you have any medicines to take, put them in the circle with you so that they become energised. Sit in the centre of the circle, with the element that represents the part of the body that you wish to heal in front of you. Focus on this. Breathe in and out deeply and slowly and relax until you reach a meditative state. With your eyes closed, now imagine a pure, white light surrounding you. Open your eyes and take the element's container in your hands. Feel the energy of the element flowing into you. Hold the element close to the area that needs healing and visualise the energy flowing into it. Imagine yourself being healed.

Continue to do this until you feel at peace. Now pinch out the candle or put it out with a drop of water and return the elements to nature: pour the water on the earth, bury the candle and crystal and pour the salt or earth outside. Now make a meal of cakes and wine or bread and fruit juice.

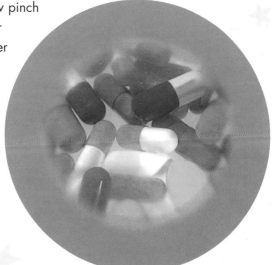

A minor ailment is thought to be eased by this ritual within a week or two and a major one within a month or two.

*Place any medicines within the circle
so that they can be energised.*

# spells for prosperity

There's no doubt about it: we all need money, and life without it is pretty intolerable. The amount that we need is debatable, as we each have different ideas about the lifestyle that we think we deserve. However, at the very least, we need to be able to pay our way and to have something left over for leisure and pleasure.

The problem with using magic to bring money is that something is always required in return, so it may be preferable to ask for a better job so that you can at least earn some extra pennies rather than receiving a large, unearned windfall.

Norse magicians used to say that the supreme god Odin was responsible for the gifts of luck, money and abundance and that he always demanded a 'gift for a gift'. There have been cases recorded where people have used magic in order to obtain large sums of money and this has worked – but in the form of compensation for an accident or injury. Worse still, even people who know nothing about magic or the spiritual world frequently find that a large win is followed by illness or even bereavement.

If you use magic to obtain money, or if it comes your way entirely by accident, you absolutely must put some of your winnings or energy into either helping others or working for the benefit of the planet. Having given you this warning, the following spells may help you out of a sticky financial situation.

# bringing money into your house

This one of many 'dumb' spells, which means that you can't talk to anybody while performing the ritual. It also has to be done early in the morning and on a new moon. My friend, Jonathan Dee, recommended this spell.

1  Gather up any small change that you have lying around the house.

2  Get dressed, but don't wash or brush your teeth.

3  Pick up the change and take it to the front door, then open the door and step outside the house.

4  Throw the change back into the house.

5  Go out to the shops and buy one or two items of food. You can, of course, start to talk once you are at the shops and also from then onwards, but you must leave the change lying where it fell for the rest of the day.

6  The following morning, pick up the change and spend it.

*Gather up any loose change
that you find in your house.*

# shoes and coins

I saw this spell written up in a modern book, but as soon as I read it I realised that it was very old and that I had come across something similar when I was a child. This spell is best performed outdoors on a sunny Monday morning, and you will need a pair of old shoes and a gold-coloured coin. In some countries, gold coins are hard to come by, but many specialist Chinese shops always stock imitation-gold coins that cost next to nothing.

Hold your gold coin in a ray of sunlight and say:

> **What I see, may it increase,**
> **so that I may have**
> **financial peace.**

Then place the coin in the left old shoe and put both shoes on. Walk clockwise in a circle three times and then remove the shoes and place them in a 'T' shape where they will not be disturbed.

Do the same on the following two days. On the third day, transfer the coin to the shoes or slippers that you most frequently wear and wear it there as often as possible for a few weeks.

It may be more comfortable to use slippers and to glue the coin to the side of one of the slippers and then to walk around inside the house rather than crippling yourself by trying to go to work with a bulky coin in your shoe. Do not spend this coin and keep it safe once you have finished with it.

# a sunday spell

Here is a more complicated prosperity spell.
It needs to be performed on a Sunday during the
period of a waxing moon.

**1** Starting on a Sunday, carry your coin around with
you for seven days. During this time, do someone a
favour or exert yourself in some way.

**2** Take a stick and draw an imaginary circle on the
floor, moving widdershins (in an anti-clockwise direction).

**3** Stand inside the circle.

**4** Use your compass to find north, south, east and
west, then sprinkle a little of the herbs on the edges of
your circle in these four directions.

**5** Stand in the middle of the four points and light your candle.

**6** Holding the candle in your right hand and the coin in your left hand, recite the spell below
while passing the coin through the flame of the candle six times:

For this, you will need the
following components:

A compass
A silver coin that you have
blessed (see page 23)
4 large pinches of herbs
1 gold candle
A piece of orange silk
around 6in (13cm) square
Orange ribbon
Fresh spearmint leaves
A stick

> *O Angel Michael, help me understand the nature of abundance,*
> *that I may become wealthy in spirit as well as financially.*
> *Bring me the money that I need in order to live comfortably.*
> *I ask for the right amount of prosperity for my needs and to allow me*
> *the energy to use my gifts to celebrate life and to help others.*
>
> *Grant me this and I will remember to give as I have received.*

**7** Thank the angel and place the coin in the silk square, along with some spearmint leaves, and tie up the lot with orange ribbon. Keep this either in your coin purse or in a box your home.

**8** Moving widdershins (anti-clockwise), gather up the herbs while saying 'thank you' to the four directions, and when you are holding all four pieces in your hand, say 'So mote it be' and visualise the spell being carried out into the universe.

**9** Place the herbs under a tree, preferably a nut tree or else an oak or a maple.

*Keep everything together at home in a box or tin.*

# moon spells

These are so well known that you may have heard of them. When you see a new moon, take a silver coin outside, look at the moon and turn the coin over and pray for financial help. Jonathan Dee says that he was always told to turn the coin over while it is still in your pocket. Another less well-known spell is always to have money in your pocket on the first full moon in the springtime. This ensures that you will have money all year.

# abundance

To some extent, we tend to focus overmuch on the idea of money, but we can ask for abundance, meaning having enough of what we need, whether this be in the form of money or not. This can be achieved by using a simple visualisation technique, which gets rid of your fears, debts, financial and other problems.

You could make your circle before doing this visualisation or could do it in a place where you feel safe and comfortable. The bath might be a good place or you could simply relax in a chair. If you wish, you could light a royal blue- or green-coloured candle and an incense stick before you start. If you want the right day, then choose a Thursday or Friday, while a perfect hour would be a Jupiter or Venus one. Peace and quiet are the only really essential ingredients for this.

Imagine that you have a stack of plastic rubbish sacks handy and imagine all of your fears and worries going into these bags. Then imagine yourself building a huge bonfire in some safe place and, once it is burning well, see yourself putting all of your bags of worries and troubles onto the fire. If you wish to aid the process with an incantation, use this one below:

*Take my worries, fears and loss*
*Take them from my shoulders now.*
*Let abundance into my life*
*So I can live in peace somehow.*
*So mote it be.*

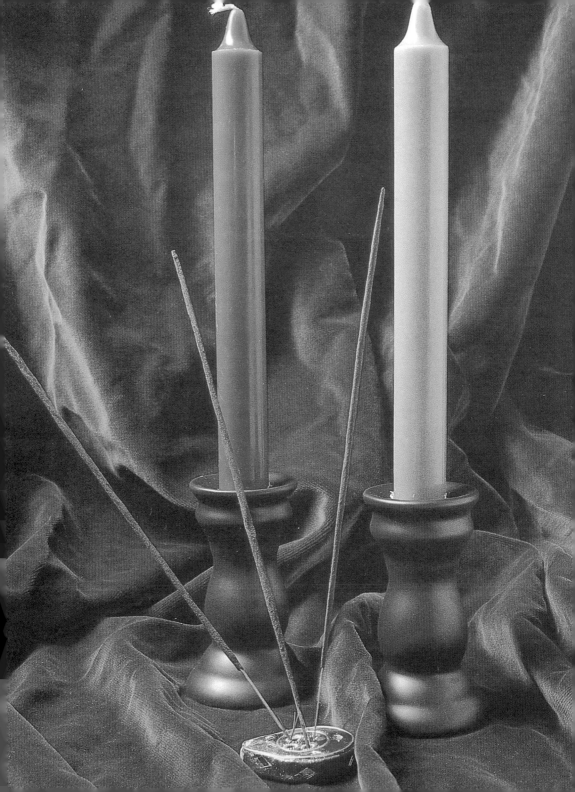

# lotteries and raffles

Buy your lottery ticket on the right day for your star sign. If you want to take this further, check out the hourly ephemeris on page 32 for the right hour to buy your ticket.

The list below will show you which planet you should be looking at in the ephemeris.

| | Sign | Day | Planet |
|---|---|---|---|
| | Aries | Tuesday | Mars |
| | Taurus | Friday | Venus |
| | Gemini | Wednesday | Mercury |
| | Cancer | Monday | The moon |
| | Leo | Sunday | The sun |
| | Virgo | Tuesday | Mercury |
| | Libra | Friday | Venus |
| | Scorpio | Tuesday | Mars |
| | Sagittarius | Thursday | Jupiter |
| | Capricorn | Saturday | Saturn |
| | Aquarius | Saturday | Saturn |
| | Pisces | Thursday | Jupiter |

# spells for all seasons and reasons

The ancient Chinese believed that gods lived on earth and returned to heaven once a year to report to the Jade Emperor on the affairs of men. Many household objects were supposed to have their own gods, who required offerings and sacrifices.

Educated people considered this to be superstition, but they kept up some of the practices just in case.

*Honey to sweeten a household god.*

# chinese spells

Oddly enough, there aren't many Western spells for a happy home, but the Chinese specialise in this area of mysticism. The first spell given here is one that invokes the hearth god.

The hearth god was – and often still is – regarded as the most important of the household gods as he saw everything that went on in the family. His report to the Jade Emperor resulted in the good and bad luck that came the family's way during the course of the following year. If you want to propitiate the household god, you must buy or make a wooden figure to represent him. At the

Chinese new year, smear his lips with honey so that he speaks sweet words to the emperor and then set off a few loud fire-crackers or bangers outside to ward off evil spirits. Finally, burn the image in a fire to send him happily on his way.

A simple form of feng shui that will ensure that the family has enough to eat during the year is to keep a small blue bowl somewhere in the kitchen with a few grains of rice in it.

*A rice bowl to keep the family fed.*

# a house of plenty

Here is an old tradition from Poland that my mother-in-law told me is designed to ensure happiness and prosperity in a new home. I have always performed this when moving house, and my daughter has also followed this superstition.

Before you bring anything at all into the new home, take in a candle, some bread and some salt. The candle should not be lighted and is simply there to ensure that you will be able to afford to heat and light the place. The bread ensures that you won't starve and the salt (as in so much European folklore) is for protection against evil.

*Salt, bread and a candle ensure good fortune in a new home.*

# a superstition for a house move

There are three months in the year during which it is not reckoned to be good to enter a new house or to sign a lease. These are April, July and November. Neither is the eleventh day of any month good for such projects.

| January | February | March |
|---------|----------|-------|
| ~~April~~ | May | June |
| ~~July~~ | August | September |
| October | ~~November~~ | December |

# a cord spell

This spell can be used to grant any wish, and is probably best performed by the kind of 'wannabe' spell-weaver who is definitely set on upward mobility! Gather together an equal number of males and females, not forgetting to provide a partner for yourself. If you do end up with an odd number, ask one person to take the part of two people and to hold two cord ends (this will become clear shortly). If you have more women than men, you

will invoke the power of a goddess rather than a god, so if the wish is for a motorbike, success for a football team or anything that is traditionally masculine, it may not be successful.

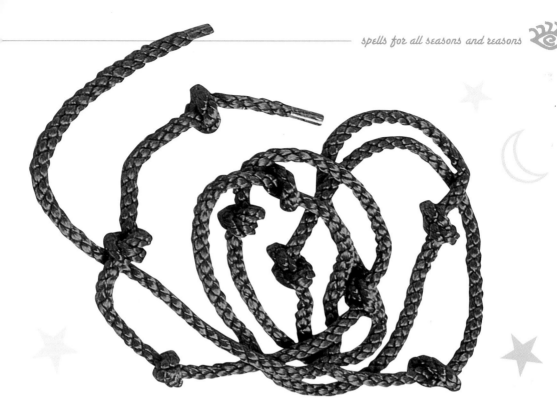

Sit your pals in a circle and hand out lengths of cord that will work out as one length per two people. The kind that is used for trimming furniture or curtains is best for this, along with some sticky tape wrapped around the ends to prevent them from fraying. Arrange your circle so that each man is sitting opposite a woman and then ask each couple to stretch one cord between them. Once all of your friends are fully equipped, ask them to manoeuvre the cords so that they all cross in one spot in the centre. Ask your circle to focus on this spot and to concentrate on the chosen wish. You, as high priestess, must then say whatever wish has been decided upon aloud, while at the same time tying a knot close to your end of the string. After you have done this, the person to your left repeats the wish and ties a knot, followed by the person on their left and so on. The wish is then spoken again and again (without further knot-tying), faster and faster around the circle in order to build up power. When you think that there is enough power, shout the word 'Drop!', at which point everybody should drop their cords.

If one or two people are alone, the following spell can be recited after stating the wish once (or once each) and tying the knot once at each end.

**By knot of one, the spell's begun.**
**By knot of two, it cometh true.**
**By knot of three, so mote it be.**
**By knot of four, the open door.**
**By knot of five, the spell's alive.**
**By knot of six, the spell is fixed.**
**By knot of seven, the stars of heaven.**
**By knot of eight, the stroke of fate.**
**By knot of nine, the thing is mine.**

# a spell for fortitude

Sit quietly and recite this prayer spell from ancient China, which calls upon the goddess of the valley.

> *The Valley Spirit never dies.*
> *It is named the Mysterious Woman.*
> *And the Doorway of the Mysterious Woman*
> *Is the base from which Heaven and Earth sprang,*
> *It is there within us all the while;*
> *Draw upon it as you will, it never runs dry.*

# a stress-buster

Sometimes you need a spell simply to give
you courage or to help you to cope with a
stressful situation. This one should do the
trick.

Make or buy a pretty bag that is large
enough to contain the elements that you
need. On a sunny day, collect some dry
earth and either collect or buy some borage
herb. If the borage needs to be dried, you
could tie it up by the stalks and hang it in
your kitchen for a while. Put the earth and
the borage into the bag and imagine some
of your thoughts, feelings and memories
going in with the ingredients. Concentrate
particularly on putting good feelings, happy
thoughts and wonderful memories in the bag.
Then tie or fasten the bag and keep it in a
safe place.

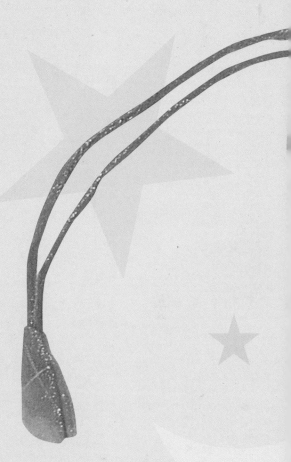

When you are feeling particularly
downhearted or stressed out, open the
bag and allow the earth, borage and
happy memories to drift around your
fingers. You should soon feel able to
cope with all that life throws at you
again.

# bereavement

Nothing on earth feels worse than losing someone you love, and there is often little in the way of practical help on hand. Time is the great healer, of course, but how do you cope until enough time has passed?

The most loving thing that you can do under these circumstances is not for yourself, but the person whom you have lost. Let them go. Ask for white light to carry them up and away to the place where they need to be. Write them a letter, perhaps, and then burn it. Say goodbye and ask your loved one to meet you again when your time comes.

Another method of achieving the same end is to light an incense-burner and put a few drops of lavender oil in it. Use a candle in the colour that links with the loved one's sign of the zodiac and then write their name on a piece of pretty paper. Using dark-blue ink, write to the person, telling them why you miss them so much. Then light the paper and burn it, saying the following:

> *From the north, the east, the south and the west,*
> *Come water, fire, earth and air.*
> *I release the spirit of [the person]*
> *To your eternal care.*

You should then pamper yourself with herbal baths, using rosemary and lavender oil and scented candles to help you to relax.

*These tools can help to ease the pain of bereavement.*

# a visualisation for breaking free

Sit in a quiet room and light candles or incense if you wish. Now imagine a figure-of-eight drawn in white on the floor. Sit in one loop of the figure-of-eight and visualise the whole design, with you in it, being flooded with healing energy. Now imagine the thing that you want to cut out of your life either sitting or placed inside the other loop of the figure-of-eight.

Finally, imagine a large pair of scissors or knife cutting through the figure-of-eight at its thinnest part. Now see the other half of the figure floating away, out of the room, and out of your life, to some far-distant place.

# a graffiti spell

Here is a very simple spell that you can do when there is something special that you want to happen that doesn't come into any of the categories that I have covered elsewhere in this book. This is best done on a new moon if possible. You can light a candle and some incense if you wish, although this is optional. Sit down quietly and write down whatever it is that you want on a piece of paper, then go over what you have written and write the same thing again in as few words as you can. (You may have to try this a few times until you can make your wish really short.)

Now take a piece of pretty paper and make a design out of all of the letters in the note that you have written, perhaps something like the kind of graffiti that you see around railway lines. Finally, burn the paper, calling on your genie or guardian angel for help. Remember to thank your genie or angel when your wish comes true.

## a comment from kenya

This one warns you not to accept blindly the harmful beliefs of others, but to draw your own conclusions. 'Do not call a person a witch before he has bewitched you.'

## an east african superstition

It is unlucky to do anything for the first time on the twenty-first day of the month.

## a magical superstition from east africa

When a person takes off a pair of shoes, if one shoe lands on top of the other, that person will set out on an unexpected journey.

# spells from south africa

Most people in the mind, body and spirit business tap into either Egyptian or Native American traditions for spiritual inspiration. Although I am quite drawn to ancient Egypt, the Native American culture has never really attracted me. I fell in love with the idea of going to South Africa when I was a little child, reading about it in my school geography book. I eventually went there, made good friends and even lived there for a while. The following spells therefore reflect my own South African influences and connections.

# something to lift your heart
# when times are hard

This amazingly modern spell, which is called the 'Song of Antuf', is taken from an ancient Egyptian tomb of around 1300 BC. Settle yourself down quietly and recite the following:

> *Increase your happy times, letting yourself go;*
> *Follow your desire and best advantage.*
> *And 'do your thing' while you are still on this earth,*
> *According to the command of your heart.*

# a starry spell for better karma

I pass on this spell courtesy of my friend, Kabbalist and all-round occult expert, Kate Rheeders, of South Africa.

This spell is designed to improve your family's karma. In the UK, you will only be able to cast it during a hot summer night because it is an overnight, outdoor spell that is best performed on a cloudless night and, if possible, during a new moon. Find a soft patch of grass and make yourself comfortable. Close your eyes for a moment, take a few deep breaths and relax your whole body, from your toes to your scalp. Create in your mind the feelings of a new-born baby that is about to open its eyes and see the world for the first time.

Now open your eyes and look at the stars. Imagine them to be loving members of your family, some still alive

and others long gone. See the way that the stars twinkle and use your imagination to find meaning in this. Pay particular attention to any meteorites or falling stars. Ask whatever question you need to ask and then wait until you feel some kind of response. Feel the answer in the area behind your navel and then allow your mind to work on it. Seek confirmation from the stars and ask them to guide and protect you.

# the rain dance

Here is another spell from Kate
Rheeders, this one being a rain spell.
We who live in Britain rarely need a
spell for rain – indeed, we are more
likely to need one to make it stop
raining – but there are places in the
world where rain is vital, so if you live
in one of those, this spell is for you.

On a cloudy day, find a patch of
grass and lie down. Allow your eyes
to scan the sky for hidden pictures in
the clouds and watch how they
change shape. Letting your intuition
flow, notice how these figures interact
with each other and, if you have
enough imagination, allow them to
tell their story. Notice how the clouds
become heavier until they leave the
safety of the sky and tumble towards
you. Get up and dance for joy. Feel
the happiness grow within you and
ask the rains to make your crops
grow into beautiful food plants.

# angels and saints

## guardian angels

Before casting any spell, you may wish to call upon your guardian angel to protect your family, your home or your friends. It is especially useful to ask your guardian angel to look out for you if you are embarking on a journey. A guardian angel might help you to find a safe parking space, protect your car from theft or prevent you from receiving a parking ticket! If you have exams to take or are participating in a sporting event, taking part in a play, cooking a special meal, attending a job interview or doing anything else that has a good chance of going wrong, call upon your guardian angel for help and protection before you set out and you will find that you should come through the ordeal with ease.

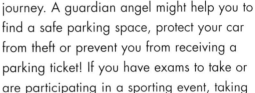

It is easy to find your own special guardian angel: just check out your birth date against the table on the right. You may notice that some guardian angels double up to serve more than one birth date – perhaps these angels have energy to spare.

| BIRTHDATE | GUARDIAN ANGEL |
|---|---|
| 1–7 JANUARY | NEHAHIAH |
| 8–10 JANUARY | YEIALEL |
| 11–15 JANUARY | HARMEL |
| 16–20 JANUARY | MITZRAEL |
| 21–25 JANUARY | UMABEL |
| 26–30 JANUARY | IAH-HEL |
| 31 JANUARY – 4 FEBRUARY | ANAVEL |
| 5–9 FEBRUARY | MEHIEL |
| 10–14 FEBRUARY | DAMABIAH |
| 15–19 FEBRUARY | MANAEL |
| 20–24 FEBRUARY | EYAEL |
| 25–29 FEBRUARY | HABUHAIM |
| 1–5 MARCH | ROCHEL |
| 6–10 MARCH | JABAMIAH |
| 11–15 MARCH | HAIAIEL |
| 16–20 MARCH | MUMIAH |
| 21–25 MARCH | VEHIA |
| 26–30 MARCH | JELIEL |
| 31 MARCH – 4 APRIL | SITAEL |
| 5–8 APRIL | ELEMIAH |
| 9–14 APRIL | MAHASIA |
| 15–20 APRIL | LELAHEL |
| 21–25 APRIL | ACHAIAH |
| 26–30 APRIL | CAHETEL |

| BIRTHDATE | GUARDIAN ANGEL |
|---|---|
| 1–5 MAY | HAZIEL |
| 6–10 MAY | ALADIAH |
| 11–15 MAY | LAUVIAH |
| 16–20 MAY | HAHIAIH |
| 21–25 MAY | IEZAIEL |
| 26–31 MAY | MABAHEL |
| 1–5 JUNE | HAZIEL |
| 6–10 JUNE | HEKAMIAH |
| 11–15 JUNE | LAUVIAH |
| 6–21 JUNE | CALIEL |
| 22–26 JUNE | LEVIAH |
| 27 JUNE – 1 JULY | PAHALIAH |
| 2–6 JULY | NECHAEL |
| 7–11 JULY | YEIAYEL |
| 12–16 JULY | MELAHEL |
| 17–22 JULY | AHEVIAH |
| 23–27 JULY | NITHAIAH |
| 28 JULY – 1 AUGUST | HAAIAH |
| 2–6 AUGUST | JERATEL |
| 7–12 AUGUST | SEHEIAH |
| 13–17 AUGUST | REIYEL |
| 18–22 AUGUST | OMAEL |
| 23–28 AUGUST | LECABEL |
| 29 AUGUST – 2 SEPTEMBER | VESARIAH |

| BIRTHDATE | GUARDIAN ANGEL |
|---|---|
| 3–7 SEPTEMBER | YEVAIAH |
| 8–12 SEPTEMBER | LEHAHIAH |
| 13–17 SEPTEMBER | CHAVAQUIAH |
| 18–23 SEPTEMBER | MENADEL |
| 24–28 SEPTEMBER | ANIEL |
| 29 SEPTEMBER – 3 OCTOBER | HAAMIAH |
| 4–8 OCTOBER | REHAEL |
| 9–13 OCTOBER | DEIAZEL |
| 14–18 OCTOBER | HAHAEL |
| 19–23 OCTOBER | MIKAEL |
| 24–28 OCTOBER | VEULIAH |
| 29 OCTOBER – 2 NOVEMBER | YELAIAH |
| 3–7 NOVEMBER | SEHALIAH |
| 8–12 NOVEMBER | ARIEL |
| 13–17 NOVEMBER | ASAIAH |
| 18–22 NOVEMBER | MIHAEL |
| 23–27 NOVEMBER | VEHUEL |
| 28 NOVEMBER – 2 DECEMBER | DANIEL |
| 3–7 DECEMBER | HAHASIAH |
| 8–12 DECEMBER | IMAMIAH |
| 13–16 DECEMBER | NANAEL |
| 17–21 DECEMBER | NITHAEL |
| 22–26 DECEMBER | MABAHIAL |
| 27–31 DECEMBER | POYEL |

# christian saints

If you prefer to call upon a strictly Christian saint, rather than ancient Jewish or Egyptian guardian angels, the following information is perfect for you. You could encourage your saint by lighting a white candle and you could also light three incense sticks.

Face towards the Holy Land, close your eyes and then pray. The following saints are said to be effective for specific purposes:

**Saint Gregoire or Saint Remi**
For job-seeking.

**Saint Anastase or Saint Benoit**
For an awkward appointment or meeting with your boss or tutor.

**Saint Estelle**
If you are hoping to get married.

**Saint Bond**
For when you want to get back together with a boyfriend or to make up with one after having had a falling out.

### Saint Agnes
If you are hoping to get married or to have a happy wedding day.

### Saint Ella
To make your boyfriend or husband less fickle.

### Saint Chronidas
For selling a house at a good price.

### Saint Monas
For courage.

### Saint Eusene
For dealing with tax affairs and other kinds of official problems.

### Saint Frunence
To stop having to spend money unnecessarily.

### Saint Barlaa
For overcoming shyness.

### Saint Francoise d'Ambroise
To stop yourself being jealous.

And finally, of course, Saint Jude, for lost causes.

# genies

Genies originated in what was once ancient Persia and is now modern Iran. Modern Iran is a Moslem country, so genies are probably no longer in fashion there – officially, at least. If you like the idea of having a genie, the following list will show you which

| BIRTHDATE | GENIE |
|---|---|
| 1–10 JANUARY | EPIMA |
| 11–20 JANUARY | HOMOTH |
| 21–29 JANUARY | OROASOER |
| 30 JANUARY – 8 FEBRUARY | ASTIRO |
| 9–18 FEBRUARY | TEPISATRAS |
| 19–29 FEBRUARY | ARCHATAPIAS |
| 1–10 MARCH | TNOPIBUI |
| 11–20 MARCH | ATEMBUI |
| 21–30 MARCH | ASSICAN |
| 31 MARCH – 9 APRIL | SENACHER |
| 10–20 APRIL | ACENTACER |
| 21–30 APRIL | ASICATH |
| 1–10 MAY | VIRAOSO |
| 11–20 MAY | AHARAPH |
| 21–31 MAY | THESOGAR |
| 1–10 JUNE | VERSUA |
| 11–21 JUNE | TEPISATOSOA |
| 22–30 JUNE | SOTHIS |

one is yours. All that you need to do to locate your genie is to consult your birth date and then you can ask your genie for help with any sticky problem.

| BIRTHDATE | GENIE |
|---|---|
| 1–10 JULY | SYTH |
| 11–22 JULY | THUIMIS |
| 23 JULY – 2 AUGUST | APHRIUMIS |
| 3–11 AUGUST | SITHACER |
| 12–22 AUGUST | PHUONISIE |
| 23 AUGUST – 2 SEPTEMBER | THUMIS |
| 3–12 SEPTEMBER | APHUT |
| 13–22 SEPTEMBER | THOPITUS |
| 23 SEPTEMBER – 2 OCTOBER | SERUCUTH |
| 3–12 OCTOBER | ATERECHINIS |
| 13–22 OCTOBER | ARPIEN |
| 23 OCTOBER – 2 NOVEMBER | SENTACER |
| 3–12 NOVEMBER | TEPISEUTH |
| 13–22 NOVEMBER | SENCINER |
| 23 NOVEMBER – 1 DECEMBER | EREGBUO |
| 2–11 DECEMBER | SAGEN |
| 12–21 DECEMBER | CHENEN |
| 22–31 DECEMBER | THEMES |

# useful herbs, trees, leaves and flowers

You can put flowers, leaves or herbs under your pillow and, when you go to bed, concentrate on the things that you need or want in the hope that the gods will go to work on your problems while you are sleeping. Some herbs can be used in cooking or brewed to make a drink, but only do this with those that are obviously safe to consume, like nutmeg, thyme, mint or commercially produced raspberry-leaf tea mix. If you are at all unsure, put the goodies under your pillow, where they can work their magic without doing any harm.

### Love, sex and courage
Basil, borage, witch hazel, **apple**, lavender, thyme, jasmine, sorrel, mint, aloe, yarrow, thorn, chestnut, cypress, hawthorn and acacia.

### Joy, fun and children
Camomile, celandine, hawthorn, hyacinth, marjoram, saffron, rosemary, ash and palm.

### Communication and improving the memory
Marjoram, **parsley**, dill, rosemary and leaves from nut trees, especially hazel.

*Sunflowers for joy, happiness and children.*

### Home life
Rosemary, comfrey, **willow**, mangrove and paperbark tree (Australia).

### Health
Marjoram, juniper, dill, nutmeg, oak and leaves from nut trees. Many ailments specific to women are eased by drinking raspberry-leaf tea. This is also an ancient, and still useful, aid to relaxation and pain relief during labour.

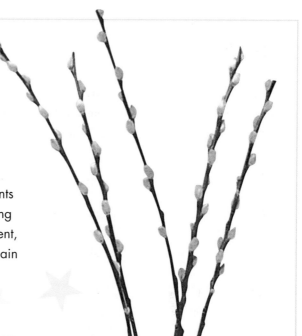

### Home life
Rosemary, comfrey, willow, mangrove and paperbark tree (Australia).

### Travel
Aniseed, marjoram, bay, clove, fennel, **pine**, oak, lime, mulberry, ash, birch.

### Protection
Sage, sorrel, bay, **cloves**, witch hazel and lime.

### Abundance and wealt
Sage, allspice, comfrey, ginger, sorrel and honeysuckle.

# flowers

Flowers have certain associations that make them useful when you want to enhance a particular spell. You may put some of the appropriate flowers in a vase and keep them near you while performing your spell.

## Love, sex, bravery and adventurousness
Honeysuckle, rose, poppy, violet, foxglove, vine, clover, rhododendron and dark-red flowers, as well as cacti.

## Communication and health
Lily of the valley, lavender, orchid, gladiolus, cornflower, small, brightly coloured flowers, snowdrop, lily and narcissus.

## Home life
Acanthus, lotus (water lily) and wild flowers.

## Joy, fun and children
Marigold, nasturtium, sunflower and cyclamen.

## Travel
Dandelion, pinks, rushes and pimpernel.

## Spirituality
Water lily and poppy.

## Protection
Sage, sorrel, bay, cloves, witch hazel and lime.

## Abundance and wealth
Sorrel, honeysuckle, pansy, ivy and thistle.

## Originality and creativity
Orchids, absinthe (wormwood) and buttercup.

*A rose for love.*

# lucky numbers

Many of us like to think that we have a lucky number or a lucky-number combination, and many people work this out for themselves as a result of observation. My husband has noticed over the years that anything he has bought that worked well for him had a number on it that added up to three, six or nine. This has been the case for houses, cars, hotel rooms and much else.

I have noticed that every time that I have lived in a house or had a car with the number five, trouble followed, despite the fact that my zodiac sign of Leo's number is five. My numerology destiny number, which is arrived at by adding together the numbers in my date of birth, is number six, and that has always been a much better number for me to work with.

## astrology

The following numbers link to each sign of the zodiac and this is particularly useful as a link to a particular person or situation that you are working upon. In this case, write the number down and wrap it around a candle of the appropriate astrological colour.

| Sign | | Number | Sign | | Number |
|------|--|--------|------|--|--------|
| ARIES | | ONE | LIBRA | | SEVEN |
| TAURUS | | TWO | SCORPIO | | EIGHT |
| GEMINI | | THREE | SAGITTARIUS | | NINE |
| CANCER | | FOUR | CAPRICORN | | TEN |
| LEO | | FIVE | AQUARIUS | | ELEVEN |
| VIRGO | | SIX | PISCES | | TWELVE |

# numerology

One can opt to cast a spell on a new moon or on the right astrological day or hour, but a nice alternative is to select the right day for a specific purpose by numerology. Simply add all of the numbers together in the date and then reduce them as follows:

21 September 2004 (9+(2+1)+(2+0+0+4) = 18).
Now reduce this to a single digit by adding the
two numbers together (1 + 8 = 9).
The number for the day is therefore 9.

## INTERPRETING THE NUMBERS

**1** STARTING SOMETHING NEW

**2** RELATIONSHIP OR PARTNERSHIP MATTERS

**3** CREATIVITY

**4** DOMESTIC OR FAMILY MATTERS

**5** COMMUNICATION, CONTACTING THE RIGHT PERSON

**6** WORK CAREER

**7** SPIRITUAL OR EMOTIONAL MATTERS

**8** MONEY, BUSINESS

**9** BRINGING SOMETHING TO A CONCLUSION, LETTING GO

# runes and tarot

If you have a set of runes, you can take one or more of them and put them into your magic circle, along with a lighted candle or anything else that you use, as a focus for what you particularly want from a spell. If you buy a new set of runes, it is a good idea to cleanse and bless them. To do this, you need a table that is clean and clear of everyday clutter. Lay a clean cloth on the table, spread out the runes in a rough circle and imagine a white light, flecked with turquoise and pale green, coming down from the universe and filling the runes with power, love and healing potential.

The same method is useful for a new deck of tarot cards or any other tool that you intend to use on a regular basis. You can even use the light to cleanse and bless a computer that you use for astrology once you have set it up.

Much of the following list of rune information was given to me many years ago by Seldiy Bate. There are slight variations in the way that rune names are spelled, and I have used Germanic types for each, in addition to their English equivalents.

### THE BLANK RUNE

This rune would not usually be used in magic because it represents fate, kismet, karma, the wheel of fortune or anything (good or bad) that is in the hands of the gods or is handed out to the enquirer by the gods.

### ETHEL, ODILA
inheritance, inherited land

This rune concerns benefit through property, gifts, help from older relatives, legacies, heirlooms, objects that have been left to the enquirer, as well as documents, wills, legal and financial matters. It also denotes inherited talents and family lore.

### FEOH, FEHU
#### cattle

Cattle represented prosperity in ancient times, much as they do in Africa today. Feoh represents the enquirer's position, funds, wealth, status and career, as well as material and emotional security and fulfilment of all kinds. This rune rules pregnancy, fertility and growth. The enquirer can use this rune to enhance his or her reputation or to bring fame.

### UR, URUZ
#### wild ox or aurochs

For physical strength, masculinity, the active principle, the physical and material plane. This rune offers opportunities for the enquirer to better themselves, and possibly also a financial improvement, but always with the expenditure of energy or strength. It also promotes the acquiring and use of physical skill, or the training of such in others.

### THORN, THURISAZ
#### thorn or giant

This rune is sacred to Thor, the Nordic thunder god, and brings protection in all of its forms, even psychic protection. It can be used to give an enquirer the strength to stand up for themselves when necessary. Don't use this rune if you want anything to happen quickly, as it is better attuned to things that take their course.

### RAD, RAIDHO
#### a wheel, chariot or the act of riding

This rune represents journeys, movement, transport, as well as learning and imagination, change, movement in the enquirer's affairs or the ability to control matters more successfully.

### OS, ANSUZ
#### a god

This rune is sacred to Odin, the chief god of magic, and represents authority, superiors, elders and ancestors, parents and all that can be inherited from them. This rune brings spiritual progress and guidance from above. It is especially useful for conjuring up sensitivity and inspiration and for anything that requires communication in all of its forms. Odin is also the god of storms, so protection from these is possible with this rune.

### KEN, KENAZ
#### a bonfire, a pitchbrand torch

This rune brings warmth and light, mental illumination and creativity, luck in celebrations and also in love, especially for a woman, plus success, status and good health.

### GYFU, GEBO
#### a gift, giving

This rune could bring a gift, a contract or an opportunity, and it can allow the enquirer to make the most of his or her talents. It brings happiness and success in all kinds of partnerships, but there is always an element of give and take, because the runes rarely give anything for nothing.

### WYN, WUNJO
#### joy

This rune is associated with a mixed bag of ideas, including joy and happiness on an emotional level, as well as journeys over, or close to, water or visitors from overseas. Artistic and spiritual awakenings, gain, luck and success due to inspiration or creativity are also indicated. Sometimes this rune brings a fair-haired man, and sometimes it means that a wish will be granted.

### HAGAL, HAGALAZ
### hail
Nobody in their right mind would use this rune for magic, as it is associated with the destructive force of a hailstorm and brings an unpleasant bolt from the blue. Even good events can turn the enquirer's world upside down, but this may be useful for an enquirer who needs his or her life shaken up.

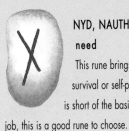

### NYD, NAUTHIZ
### need
This rune brings whatever is necessary for survival or self-preservation. If the enquirer is short of the basic necessities or in need of a job, this is a good rune to choose.

### IS, ISA
### ice
Like Hagal, this is another rune that nobody would choose as it freezes everything, including one's finances. About the only use that it might have is to induce patience or perhaps to preserve a status quo.

### GER, YER, JERA
### cycle or harvest
This is a good rune to choose if the enquirer wants to end a cycle and clear the decks. It helps with such things as being able to clear debts or get rid of unwanted baggage and is also associated with rewards for previous work.

### EOH, EEWAZ
### yew tree or yew bow
The first idea here is to use this rune for an enquirer who needs to take a flexible approach to life, to become more adaptable or to turn an inconvenient situation to an advantage. The second is a wish for long life.

### PEORTH, PERTHRO
### the secret Rune,
### a dice-cup
This rune is for those who seek knowledge or education or who want to increase or improve their psychic powers. It is also a 'lucky-break' rune.

### EOLHS, ELHAZ
### a hand held up in greeting,
### a reed, protection
This rune brings artistry, creativity, poetry, literacy and success in crafts and hobbies. It may also mean joining a club or group that has a special vocabulary. It can protect or defend against danger.

### SYGEL, SOWILO
### the sun
Hitler pinched this rune for the Nazi SS because he knew it as a symbol of success. Well, apart from the urge to start an organisation for genocidal maniacs, your enquirer can use it to bring optimism, success and victory in undertakings or in sport. It also augurs well for fame and the recognition of talent.

### TIR, TYR, TIWAZ
### the god of war, justice

This rune represents justice, the ability to stand up for oneself and even to go on the attack when necessary. It can be used to bring a love affair into being, but this would be the hot kind, filled with passion, but also aggressive arguments.

### BEORC, BERKANA
### birch tree

A fresh start, new beginnings, expansion, awakening, growth and fertility. This is a great rune for joy in the family, as well as weddings and other celebrations.

### EOW, EHWAZ
### a horse

This rune concerns travel and transport and also the media for putting across ideas to others. It implies the trust that a horse and rider share, and thus trustworthy partnership activities of all kinds.

### MAN, MANNAZ
### man or humankind

This rune represents a male figure, an authority figure or a professional man. If the enquirer needs the help of a professional or an expert, this rune will bring the right person along. It also represents oaths and agreements.

### LAGU, LAGUZ
### lake or water

This feminine rune is connected to the moon goddess. It aids change and helps an enquirer who is going into unknown territory or is about to do something that is new to him or her. It is particularly associated with women's health, fertility and childbirth.

### ING, INGWAZ, INGUZ
### the Danes,
### the god Ing

This rune has a great deal to do with keeping, breeding and maintaining cattle, which makes it, along with the first rune, Feoh, of particular interest to farmers. However, it brings fertility, production, material results and fruition in other fields as well. It can even be associated with the state of the earth itself. Use this rune for the completion of projects or dealing with something on a long-term basis and for problem-solving. This feminine rune can be used for sexual matters, female problems and children.

### DAEG, DAGAZ
### day

This rune is associated with success in studies and examinations, as well as changes for the better and success generally. It also signifies anything that is not secret and can be discussed or performed out in the open.

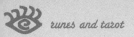
# tarot cards

Despite the fact that I use the Tarot extensively, I have not made much use of it in this book. This is because true Tarot magic is rather unpleasant and best left to the realm of horror films. However, you could use the major arcana of the Tarot by picking a card that represents what you or your enquirer want and using it as a focus. Take the card into your magic circle, along with candles or any other tools that you normally use.

There are lots of books on the meanings of the Tarot, some of which are very deep and involved, but here is a very abbreviated list that you may find useful in the meantime.

# 0 – THE FOOL
New beginnings and going into something that is completely new, of which the enquirer has no prior knowledge or experience.

# I – THE MAGICIAN
Using skills or salesmanship in a new venture. Putting known skills to work in a new way. A new job, self-employment and an important man are all possibilities with this card.

# II - THE HIGH PRIESTESS
Gaining occult knowledge or a normal education and finding the right teacher for the job. Using logic and intuition, rather than emotion, to solve a problem.

# III – THE EMPRESS
Fertility, fruitfulness and things that bring the desired result. Pregnancy and motherhood or anything concerning a mother figure.

# IV – THE EMPEROR
Being in charge of anything, even simply being in charge of one's own environment or being able to cope with difficulties. Success in business. An important man.

# V – THE HIEROPHANT
Spiritual guidance and finding a teacher or spiritual helper. Also traditional ways, or the right ways of doing things, avoiding

*The lovers.*

criminal or underhand activities.

# VI – THE LOVERS
Love, partnership, beauty and creativity.

# VII – THE CHARIOT
Travel, movement, vehicles and transport. Also victory in a difficult situation or the ability to control conflicting demands.

# VIII – STRENGTH
Health, vigour and strength. The ability to put up with something or to handle difficult situations with tact.

The hermit and the wheel of fortune.

### XII – THE HANGED MAN

One wouldn't normally choose this card for magic as it implies understanding or initiation into a new situation via pain or sacrifice. Suspension, waiting for things to come around.

### XIII – DEATH

If you want to bring something to a complete end, choose this card.

### XIV – TEMPERANCE

Peace, moderation and everything working in the right way and in the right order.

### IX – THE HERMIT

Peace and quiet, time to retreat from life and to reflect on things. Time and space to learn whatever needs to be learned. Enlightenment after studying.

### X – THE WHEEL OF FORTUNE

Choose this card when a real change is needed.

### XI – JUSTICE

Justice, fair play and help with legal matters. Creating balance in one's life.

The hanged man.

## IX – THE HERMIT

Peace and quiet, time to retreat from life and to reflect on things. Time and space to learn whatever needs to be learned. Enlightenment after studying.

## X – THE WHEEL OF FORTUNE

Choose this card when a real change is needed.

## XI – JUSTICE

Justice, fair play and help with legal matters. Creating balance in one's life.

*Judgement and the world.*

*The tower and the star.*

## XII – THE HANGED MAN

One wouldn't normally choose this card for magic as it implies understanding or initiation into a new situation via pain or sacrifice. Suspension, waiting for things to come around.

## XIII – DEATH

If you want to bring something to a complete end, choose this card.

## XIV – TEMPERANCE

Peace, moderation and everything working in the right way and in the right order.

# roman gods and symbols

You may wish to invoke a particular god for a particular purpose. For instance, you may want a candle, paper, ink, or even a talisman, in an appropriate colour. You may wish to choose a specific hour in which to make your spell so as to tap into the energy of the god in question. Alternatively, you could choose a charm, or a picture of the astrological symbol, as a focus for your thoughts.

The list below shows you who to invoke for each purpose. Don't forget to use the appropriate colours and symbolism when making spells that involve people – simply link the colours or objects to their star sign.

## MARS – ARIES – the Ram
### red

Mars was once the Roman god of war and has always been seen as aggressive and dynamic. The actual planet has a slightly reddish colour. The ram first appeared in ancient Egypt, representing the first of the gods, Ammon. This is very apt for the first sign of the zodiac. In ancient times, Aries was also symbolised by the goose, although the origins of that connection have been lost in the mists of time. Invoke Mars or the ram when you want to do something adventurous or something that takes effort and courage, especially competitive sports.

## VENUS – TAURUS – the Bull
### pink or green

Venus is associated with luxury and such activities as eating, drinking and merrymaking. This is also the planet of love. It was in the shape of a bull that Jupiter kidnapped Europa and founded the Cretan culture. The Cretans forever afterwards worshipped the bull, calling it the earth-shaker. The bull is a symbol of strength, endurance and sexuality. Invoke Venus or the bull when you need strength and endurance, when you are asking for help in matters of love or else in matters of personal funds or goods of value.

## MERCURY – GEMINI –
### the Twins
### yellow

The Roman god Mercury was the quick-witted messenger of the gods. The symbol of the heavenly twins (Castor and Pollux) is known in many cultures, including the Native American one. The Bible emphasises the rivalry of twins in the story of Cain and Abel.

Invoke Mercury or the twins, Castor and Pollux, for all magical reasons and when travel or communication are what is wanted.

## THE SUN – LEO –
### the Lion
### gold or orange

The sun is associated with the Roman god Apollo, who was the god of music, poetry, prophecy, reason, light and healing. This sign commemorates the first of the labours of Hercules, which was the defeat of the Nemean lion. The lion has always been associated with strong men, such as Samson.

Invoke Apollo or the lion for success in endeavours, business, showbusiness-type events or for matters relating to children.

## MERCURY – VIRGO –
### the Virgin
### yellow

Mercury, the quick-witted messenger of the gods, was also the god of healing and magic, as well as thieves.

This sign represents the mother goddess who presided over the gathering in of food. To think of Virgo as a prudish old maid is a mistake, because she is only one of the aspects of the great goddess or the mother-figures of mythology. Perhaps this is a girl who is to be a mother at some future date, or maybe she is a particularly loving big sister? This sign is also associated with the harvest, as well as the storage of foodstuffs.

Invoke Mercury or the harvest maid for healing or a better job or perhaps to ask for a good harvest.

## THE MOON – CANCER –
### the Crab
### silver, white or mother of pearl

The moon represents the feminine principle and is connected to the ideas of travel and the sea. The moon is associated with the goddess Persephone, also sometimes known as Selene, along with Diana, the virgin huntress and twin sister of Apollo. The crab has a hard shell and a soft interior; it goes about things moving in a sideways direction. Legend says that the crab was thrown into the sky for pinching Hercules on the toe while he was involved in one of his difficult labours. Invoke the moon or the crab for all emotional issues, in addition to household and family matters.

## VENUS – LIBRA –
## the Scales
## pink or green

Venus was the Roman goddess of love and romance. Venus is also associated with luxury and eating, drinking and merrymaking.

The poet Manilius, writing in the first century AD, said of this sign: 'Day and night are weighed in Libra's scales,/Equal a while, at last the night prevails'. This suggests the order of light and darkness.

The scales obviously have something to do with weighing and measuring and are also a symbol of the legal profession, along with the sword of justice. Librans like to weigh up all sides of an argument before making a decision.

Invoke Venus when you are looking for balance, fair play, love and harmony, as well as peace of mind.

## PLUTO AND MARS – SCORPIO –
## the Scorpion or the Eagle
## purple or magenta

Pluto, the Roman god of the underworld, is associated with great wealth, most of which is hidden underground, as well as

birth, death and sex. Before the plant Pluto was discovered, Scorpio was assigned to Mars, the red planet named after the Roman god of war. The scorpion once made the mistake of stinging the giant Orion, who threw him into the sky as far away as possible, where he still rests. The scorpion turns its sting onto itself if it feels threatened. The ancient symbol for this sign used to be the eagle. This denotes the higher side of Scorpio, which can soar above everything that is base, coarse and petty.

Invoke Pluto or Mars, the scorpion or the eagle, when you need to cope with great changes or when you have to deal with others regarding a financial or business matter.

## JUPITER – SAGITTARIUS–
## the Centaur or the Archer
## royal blue

Jupiter (Jove) was the king of the Roman gods. He was jolly and jovial most of the time, but when he was in a temper he hurled thunderbolts. The king of the centaurs was a noted archer called Chiron, who taught the Greek heroes Hercules, Theseus and Jason not only how to shoot straight, but also how to heal. Jupiter placed Chiron among the stars following an accident that left him wounded and unable to heal himself.

Invoke Jupiter, the centaur or the archer when you intend to travel or take exams or an educational course or else for spiritual enlightenment, health or luck.

## NEPTUNE AND JUPITER – PISCES –
### the Fish
### sea blue-green

The Roman god Neptune
was said to rule the sea
and all of the mysterious
things that are hidden from sight under it. Before Neptune was
discovered, Pisces was said to be ruled by Jupiter. Two fish that
are tied together and are swimming in different directions are
the symbol of Pisces. The Babylonians knew the constellation of
Pisces as 'Kun', which means 'the tails'. This is very appropriate
for the last sign of the zodiac. It was also known as the leash,
which was the link between the two fish. The symbol itself may
commemorate the occasion when Venus and Cupid disguised
themselves as fish in order to escape from the angry giant.

Invoke Neptune or the two fishes when you embark upon any
kind of paranormal or mystical experience. It is also auspicious
for artistic or musical pursuits and for sport – especially water
sports – and travel over water.

## URANUS AND SATURN – AQUARIUS –
### the Water Carrier
### day-glo colours, pale blue, brown or grey

The planet Uranus behaves very differently from all of the others in
the solar system and therefore represents unpredictable behaviour
and eccentricity. Before the discovery of Uranus, Saturn was
considered to be the ruler of Aquarius. Beginners in astrology are
often confused because although the symbol of Aquarius is a water
carrier, it is actually an air sign. The water carrier represents a
reservoir of knowledge that is hidden in the water. Other aspects
of this profound symbol include a cloud of water carried through
the air. This is a frightening symbol for people of the post-World
War II generation because it carries images of nuclear fallout.
Uranium is associated with Aquarius, too.

Invoke Uranus or the water
carrier when you want to get
an original idea or an invention
off the ground or if you want to
make new friends.

## SATURN – CAPRICORN –
### the Goat
### brown or grey

Saturn, a rather
gloomy Roman
god,
represents
limitations and hard lessons in life, but the musical and
cheerful god Pan is also linked with Capricorn, as are long-
term enterprises.

This sign is said to represent the mythical goat Amalthea,
who suckled Jupiter in his infancy. Playfully, the baby god
pulled off one of the goat's horns, which then became the
cornucopia, or horn of plenty. In addition, the mountain
goat climbs onwards and upwards in a sure-footed
Capricornian manner.

Invoke Pan when you wish to perform magic or Saturn and
the goat if you need to talk to authority figures, ask for a
raise or cope with a difficult or long-term project.

ismans*

# talismans

There are so many talismans, amulets, charms and mascots that it wouldn't be difficult to produce a large book on these in its own right. However, here are a few that I have chosen as a representative selection from several different traditions.

## CHRISTIAN SYMBOLS

The most recent of the talismans that I have chosen is the Christian cross. The fact that a person could suffer crucifixion and still keep his faith suggested that this was a faith worth keeping, which gives the cross its strong spiritual significance. This is even more so when worn by Catholics in the form of a crucifix. The symbol of the cross only came into use some three or four hundred years after the death of Christ, and in earlier times the fish was the Christian symbol, being a reminder of the fact that Christ called himself the 'fisher of men'. I doubt whether anyone would consider wearing a cross for luck, because it is still considered a reminder of faith, as is the fish. Another Christian symbol is that of St Christopher, who supposedly carried the baby Jesus to safety across a river. This is so connected to the idea of safe travel that it turns up on key rings that are sold in car-spares shops.

## THE STAR OF DAVID

Being Jewish, I often wear a six-pointed star of David on a chain necklace. The star is made up of two triangles which in turn represent the trinity that is inherent in many religious beliefs, examples of which are given below:

**Braham, Vishnu and Shiva.**
**Osiris, Isis and Horus.**
**The Holy Trinity.**
**Heaven, earth, humanity**

The two triangles are said by astrologers to mean 'as above, so below'. Astrologers and spiritual people like the star of David because it can't be confused with the five-pointed star favoured by magicians and occultists. This emblem was never a mystical symbol to Jews, being merely the logo that was painted on the shields carried by King David's troops during their many battles. Jews still enjoy wearing the star of David, as do many people who follow a spiritual lifestyle or are astrologers.

## CROSSING FINGERS AND TOUCHING WOOD

Both crossing one's fingers and touching wood relate back to the days when Christianity was an underground religion and these were secret, symbolic actions that must have been along the lines of the Masonic handshake. Crossing the fingers was symbolic of making the sign of the cross, while touching wood symbolised touching the cross.

Another, much more common, Jewish incantation against evil is the Yiddish ken ayn ahora, meaning, 'let no evil eye fall upon it'. If a visitor commented that his host's child was growing strong and tall, the child's mother would immediately respond with a fast ken ayn ahora. The evil eye was supposed to emanate from the bad wishes of those who were envious of something that one had – even a healthy child. Well, any spiritual healer will tell you that good and bad thoughts contain power, so perhaps there is a value in this.

## THE ANKH

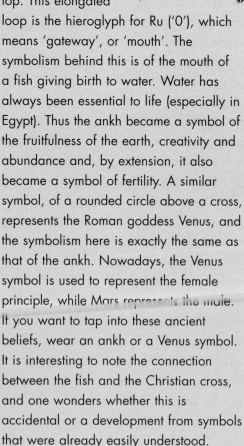

An easily recognised symbol is the ancient Egyptian ankh, which is similar to a cross, but with a loop at the top. This elongated loop is the hieroglyph for Ru ('0'), which means 'gateway', or 'mouth'. The symbolism behind this is of the mouth of a fish giving birth to water. Water has always been essential to life (especially in Egypt). Thus the ankh became a symbol of the fruitfulness of the earth, creativity and abundance and, by extension, it also became a symbol of fertility. A similar symbol, of a rounded circle above a cross, represents the Roman goddess Venus, and the symbolism here is exactly the same as that of the ankh. Nowadays, the Venus symbol is used to represent the female principle, while Mars represents the male. It you want to tap into these ancient beliefs, wear an ankh or a Venus symbol. It is interesting to note the connection between the fish and the Christian cross, and one wonders whether this is accidental or a development from symbols that were already easily understood.

## THE SWASTIKA

The oldest-known talismanic symbol is the swastika, which has been found engraved on stone implements from the Neolithic period. The legs of the swastika have been turned both ways at various times. In India today, it is usually seen with the legs pointing clockwise, whereas the infamous Nazi symbol had its legs pointing anti-clockwise.

In Sanskrit, the symbol means happiness, pleasure and good luck. The swastika is still used all over the Orient as a lucky talisman. Hitler was interested in mystical symbols, hence his use of the swastika.

## THE AXE

Another ancient contender is the double-headed axe today favoured by bikers. The axe represented the strength of a tool that made heavy work possible in ancient times.

## THE CROSS OF ST BENEDICT

Coming back to the idea of spells, the cross of St Benedict was once a popular charm that was worn as a protection against disease and the danger of being seduced into wrong-doing by wicked people. Each letter stands for a word in a piece of Latin text that was turned into a kind of mantra spell.

*Crux Sancti Patris Benedicte.*
*Crux Sancta sit mihi lux.*
*Ne daemon sit mihi dux.*

*Vade retro Satana,*
*Ne suade mihi vana;Sunt mala quae libas,*
*Ipse venena bibas.*

*Cross of the Holy Father Benedict.*
*Holy Cross be my light.*
*Let no evil spirit be my guide.*
*Get thee behind me Satan,*
*Suggest no vain delusions;*
*What thou offerest is evil,*
*Thou thyself drinkest poison.*

## A CHARMED RING

If you fancy wearing an ancient charm, have a ring made up with an image of your own head carved into a piece of green jasper. Then have the following initials –

*B.B.P.P.N.E.N.A.*

– engraved around the inside of the ring. This is an ancient Greek lucky charm, and the initials stand for 'wear this and thou shalt in no wise perish'.

## THE TALISMAN OF VENUS

Here is a talisman that has everything crowded on it and is usually worn for success, good fortune in love, joy and good fortune while travelling. The inscription says:

*Accipe my petitione, O domine, keep*
*me as apple of an eye, hide me under*
*the shadow of thy wings from all evel,*
*up Lord and help us for thou art my*
*strong rock and my castle. Amen.*

## THE FIVE EMPERORS' TALISMAN

Believe it or not, you can actually buy this talisman in any specialist Chinese shop. If you do buy one, cleanse it before using it by leaving it out in natural sunlight for a few hours. Then hold it in your hands, visualise white light coming down onto it and ask your spiritual guide or guardian angel to bless it for you. If you make your own talisman, rinse most of the component parts in a natural source of water, such as collected rain water or bottled spring water, and dry them carefully on a piece of kitchen paper before you assemble them. To make this talisman you will need a little sword or paper knife, five Chinese coins (these have holes in the middle) and some red string. String the coins onto the knife in a pattern of two, one beneath the other, then one on its own, and then two again. This can then be hung up somewhere safe.

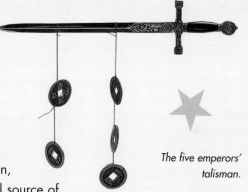

*The five emperors' talisman.*

## A CHINESE LOCKET

A small silver lock on a necklace is an ancient lucky symbol. This was originally made from of a few silver coins that had been melted down and is traditionally given to a young boy to wear. It was believed to preserve the boy from evil spirits, lock him into life and act as an aid to health and longevity.

## THE PA KUA

The Pa Kua, or eight-sided Chinese talisman that you see in books on feng shui, is a powerful method of keeping evil spirits out of your house. Hang this on your front door or place it somewhere so that it looks outwards from the front of your house.

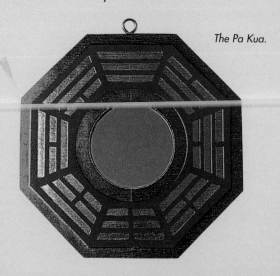

*The Pa Kua.*

## ABRACADABRA

This amusing word, so beloved of magicians at children's parties, was originally considered to be a powerful talisman and also a form of spell. It was written in one of two distinctive shapes and worn in some kind of container around the neck. The usual method was to wear the charm for nine days and then to take the talisman out of its container and throw it over the left shoulder into a running stream. The meaning of the charm, translated from the Hebrew is, 'pronounce the blessing'.

ABRACADABRA
BRACADABR
RACADAB
ACADA
CAD
A

ABRACADABRA
ABRACADABR
ABRACADAB
ABRACADA
ABRACAD
ABRACA
ABRAC
ABRA
ABR
AB
A

*The word 'Abracadabra' is a form of spell.*

## AMULETS AND MASCOTS

We all love the lucky charms that are sold as lockets or necklaces. Years ago, charm bracelets used to be very popular, but what do those charms mean, and why are they lucky? The following list covers a few of the more popular charms and shows what you might achieve if you buy and wear them or give one to a friend to wear.

To make your charm effective, hold it in both of your hands, close your eyes and imagine that you are bringing down white light from heaven, the universe or outer space. Ask your charm to give you protection and to help you with whatever project is closest to your heart. Now you can wear your charm or, if it is for someone else, you should keep it wrapped in a piece of clean paper or cloth until you can give it to your friend.

## BEANS

Beans drive away evil influences, protect children and promote happiness. Believe it or not, there is a variety of bean or nut called a malluka bean, which is normally white. In Scotland, this used to be strung around a child's neck, and if anyone sent bad vibes to the child, the bean would turn dark brown or black. Dried beans or peas were put into Christmas puddings before coins and thimbles were used. The person who found the bean was called the 'king or queen of revels' and was said to be in for a fortunate year. The following poem was recited:

*Now, the mirth comes*
*With a cake full of plums,*
*Where bean's the kind of sport here,*
*we must know*
*The pea also*
*Must revel of the queen of the court here.*

## BELL

This is one of the oldest charms in the world. Children were placed near bells to keep danger at bay and for an improvement in health. Bells were put on horses, and later bikes, for luck and safety while travelling rather than as a warning for people to get out of the way.

## BROOMSTICK

Some gypsies say that if an ill-wisher approaches your house, you should place a broomstick across the threshold in a slightly offset manner so that a cross is made. If the enemy enters, he will be powerless.

*A lucky cat.*

## BUSINESS

For success in business, carry any one of the following charms: an amethyst, bee, bull, cornucopia, cricket, deer, dragon, grasshopper, elephant, fish, fox, hammer, moonstone, narita stone or olivine stone.

## CADUCEUS

Mecury's rod of healing and peace, which was used for settling quarrels, driving away sickness and giving eloquence and youthfulness, was called the caduceus. The original rod and wings were said to have been given to Mercury by Apollo. One day, Mercury saw two snakes hissing at each other about to engage in a deadly fight, so he decided to try out the rod's power by laying it between the fighting snakes and, lo and behold, the snakes went peacefully on their way. Since that time, two snakes have been included in the symbol so that humans may rid themselves of their lower nature and rise to greater spiritual heights.

## CAT

All cats are considered lucky, even black ones, as long as they are treated kindly. If a cat comes to you and asks for your help, you must give it, or bad luck will surely follow. Cats, of course, were considered to be witches' familiars, so upsetting a cat would definitely not be a good thing to do.

## CLOVER

Clovers bring luck in general and also success.

*One leaf for fame,*
*And one leaf for wealth,*
*And one for a faithful lover,*
*And one to bring you glorious health*
*Are all in a four-leaf clover.*

## CONFETTI

Confetti has gone through a few changes in its time, and now it would perhaps be a good idea to return to its original form. It started life in Saxon times as red and white rose petals, representing both passion and purity. The Chinese used rice at weddings as a symbol of wealth and fertility, and this was later taken up by Europeans. Eventually, rice fell out of favour because it was considered wasteful and potentially dangerous (because it attracts mice and rats). Paper confetti has been fashionable for many years now, but many people are turning back to using rice because this leaves less mess than confetti and is easier to sweep up, especially on a wet day. Rose petals would be even nicer.

## DAISY

The ancient idea of pulling petals from a daisy and reciting 'he loves me, he loves me not' comes from the belief that daisies come under the rulership of Venus, the goddess of love, and are symbolic of Loyalty in love. If a friend gives a person a bunch of daisies, this is supposed to bring happiness. If your lover has left you, sleep with a daisy root under your pillow, and he will come back.

## DISC

A disc brings luck, particularly in games and sports.

## DRAGON

A Chinese charm for happiness and long life.

*A dragon symbolises a long life and happiness.*

## EYE

An eye wards off evil spirits, brings clairvoyance and clear speaking and writing. It is a favoured charm in the Middle East.

## FANG

A fang denotes strength, fertility and protection from accidents and injury.

## ELEPHANT

An elephant gives success in business and at work.

## FAN

For health and happiness, choose a fan.

## FISH

A fish will help you to realise your dreams. It is also an ancient Christian symbol.

## FROG

For recovery from illness, or to win love and friendship, obtain a frog talisman.

## HAND

Seen all around the Middle East, the hand is a symbol of Lady Fatima, a favourite daughter of the prophet Mohammed. It represents generosity, hospitality and divine goodness.

## HEATHER

A sprig of white heather is supposed to protect against danger and to grant wishes.

## HORSESHOE

The horseshoe has a very ancient history. Originally denoting the crescent of Isis and made of iron, it was supposed to guard against the evil eye. Greeks and Romans nailed horseshoes on the doors of their houses as a charm against the plague.

*A fish talisman will help you to realise your dreams.*

## HOT CROSS BUNS

If you keep a hot cross bun from one year to the next, your house will not burn down.

## MAGNET

For attracting luck and also for good health.

## MAZZUZAH

This is a popular Jewish prayer charm that can be bought from any shop that specialises in Jewish artefacts. One type of traditional mazzuzah is a slim, decorated container that is screwed to the jamb of a door. Many Jews will fix this to the front door, while ultra-orthodox Jews will fix one to every door jamb inside and outside the house. The little container holds a piece of parchment or paper with a prayer for luck and protection inside it. When I was young, tiny mazzuzahs were worn as necklace pendants, complete with the paper inscription. Alternatively, a small square of gold, with Hebrew lettering cut through it spelling mazel was worn, but these seem to have gone out of fashion and it is more popular either to wear a star of David or the Hebrew letter Hai, which means 'life'. The word mazeltov is often said in place of the word 'congratulations', or when wishing someone good luck. It is generally thought to mean 'good luck', but doesn't mean precisely that at all. The word tov is Hebrew for 'day', and mazel means roughly 'may the stars be in a good alignment for you', so it is, in a way, like asking for one's horoscope to be on one's side when something important is about to be embarked upon.

## KEY

A key talisman is for love to last and for projects and ideas to come to fruition. It is so ancient that it goes right back to Neolithic times and was used to guard the doorway of a hut against evil entering. Three keys unlock the doors to love, health and wealth.

## MISTLETOE

Although attributed to the druids, this is a Norse charm that was said to be good for lovers. Mistletoe should never be carried into any sacred building or used in church decorations, probably due to its pagan origins.

## PADLOCK

A padlock brings health and also happiness. The Chinese give silver lockets to children as a protection from disease and for longevity.

## PEARL

I was always told that pearls are for tears, the origin of this belief being due to the pearl being the jewel of Isis, who certainly suffered enough. Despite this, we all love to wear pearls. Eastern divers carried pearls as a charm against sharks. The Chinese use ground pearl for stomach upsets and the Romans used them as a cure for lunacy.

*A padlock brings health and happiness.*

## PENTAGRAM

The five-pointed pentagram represents earth, air, fire, water and spirit. It is worn by white witches as a religious symbol and also to protect them against evil.

## RING

A ring symbolises eternal love and marriage.

*The pentagram is a traditional magical symbol.*

## SALT

Salt is considered to be a powerful agent against black magic, bad vibes and danger from the jealousy and hatred of others. Put a little in a silver pendant and wear it around your neck.

*Salt is powerful talisman against danger.*

## SCISSORS

Use scissors to cut away bad influences or anything else that you need to be rid of.

## SHAMROCK

The lucky, four-leafed shamrock derives from druidic times and was probably thought to be a lucky find because it is so rare. The normal, three-leafed shamrock represents the Holy Trinity. When St Patrick was bringing Christianity to Ireland, he used a shamrock to illustrate the idea of 'three in one'.

## SHOE

A shoe denotes protection against enemies.

*Scissors help you to cut the bad things out of your life.*

## SNAKE

A snake is for a long life and wisdom, as well as good health and a great sex life.

## SPIDER

The Romans regarded the spider as a lucky symbol, especially for business or trading.

## STAR

Humans have always viewed stars as good-luck symbols. Even anti-religious communist Russia and China used a star as a national symbol.

## TORTOISE

For a long life and protection against bad people or being drawn into doing evil.

*A tortoise brings long life.*

## UMBRELLA

The umbrella in its present form was perfected by an Italian immigrant to England who lived in Hanway Street, just off Tottenham Court Road, in London, and was sick of being drenched by the downpours of that rainy city. Parasols were used before umbrellas to keep the sun off people's faces, as it was once considered bad form for a lady to allow her skin to become sunburned. Parasols have been used for centuries in the Orient, and the umbrella is a Hindu mascot that is also venerated as one of the 'eight glorious emblems of Buddha' that bring great fortune with them.

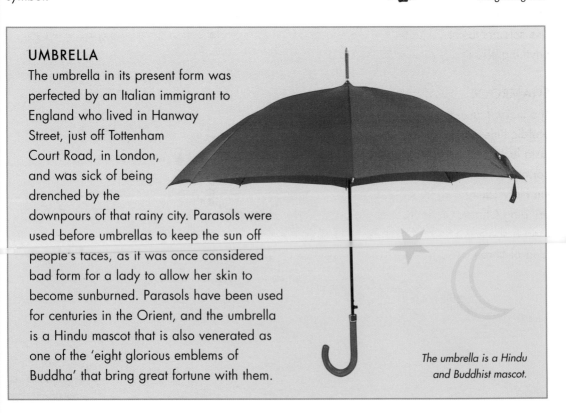

*The umbrella is a Hindu and Buddhist mascot.*

## WHEEL

Wear a wheel emblem when your luck needs to be changed (or for the wheel of fortune to turn in your favour) and also for success in law suits. In Roman mythology, this was the symbol of the goddess Fortuna. The following is part of a song immortalised in the oratorio Carmina Burana by Carl Orff. It is from an ancient Latin text that was discovered some time early last century in Germany.

> *Fortune rota volvitur:*
> *descendo minoratus;*
> *alter in altum tillitur;*
> *animis exaltatus.*
> *'The wheel of fortune spins:*
> *one man is abased by its descent;*
> *the other is carried aloft;*
> *all too exalted.'*

*The zodiac.*

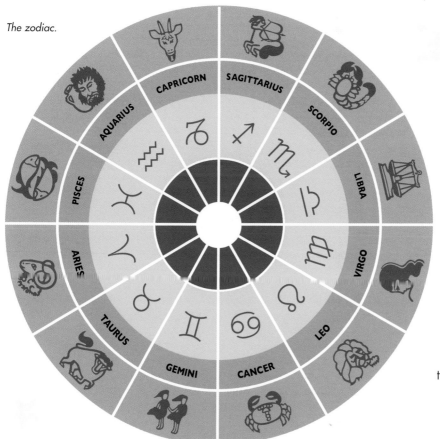

## ZODIAC

Wear your sign of the zodiac in order to invoke the god or goddess that is associated with it.

# bibliography

**The Book of Talismans, Amulets and Zodiacal Gems**
William Thomas and Kate Pavitt
Bracken Books
Published in 1993, but obviously taken from a much older manuscript.

**The Good Luck Book**
Elizabeth Villiers
Senate, 1994, first published in 1923.

**Spells and How They Work**
Janet and Steward Farrar
Hale, 1990.

**Moon Signs**
Sasha Fenton
Aquarian Press, 1987.

**Tarot in Action!**
Sasha Fenton
Aquarian Press, 1987.

**The Chinese Pakua**
Ong Hean-Tatt
Pelanduk Publications, 1991.

**Shona Customs**
Edited by Clive and Peggy Kileff
Mambo Press, 1970.

**Some Traditional African Beliefs**
Kate Rheeders
Headway, Hodder & Stoughton, 1998.

**African Proverbs and Wisdom**
Julia Stewart
Citadel Press, 1997.

**Tao Te Ching**
Lao Tzu
Wordsworth Editions Ltd
This edition 1997, but the text is centuries old.

**Dictionary of Earth Mysteries**
Janet and Colin Bord
Thorsons, 1996.

**West Country Witchcraft**
Roy and Ursula Radford
Peninsula Press, 1998.

**Principles of Paganism**
Vivianne Crowley
Thorsons, 1996.

**The Perpetual Almanack of Folklore**
Charles Kightly
Thames and Hudson, 1994.

**Llewellyn's Witches Datebook 1999**
Llewellyn Publications, 1998.

**Incense**
Leo Vinci
Aquarian, 1980.

**Love Magic**
Laurie Cabot with Tom Cowan
Souvenir Press Ltd, 1992.

**Traditional Chinese Medicine**
Rey Tiquia
Choice Books, 1996.

**Notions and Potions**
Susan Bowes
Thorsons, 1997.

**Tokolosh, African Folktales Retold**
Diana Pitcher, illustrated by Meg Rutherford
Tricycle Press, 1980.

**Spellweaving**
Sally Morningstar
Select Editions, 1995.

**Celtic Mysticism**
Anthony Duncan
Select Editions, 2000.

**Chinese Gods and Myths**
Various authors
Quantum Books Ltd, 1998.

**Aphrodisiacs and Love Magic**
Pamela Allerdice
Prism Press, 1989.

**Spells for Teenage Witches**
Marian Baker
Kyle Cathie Ltd, 2000.

**The Book of Spells**
Nicola de Pulford
Simon and Schuster, 1998.

**Llewellyn's 1999 Magic Almanac**
Llewellyn Publications, 1998.

**Advanced Magical Arts**
R J Stewart
Element, 1998.

**The Hand of Destiny**
C J S Thompson
Rider & Company Ltd, 1995,

**Occult in the West**
Michael Williams
Bossiny Books, 1979.

**Voodoo and Magic Practices**
Jean Kerboull, translated from the French by John Shaw
Barrie and Jenkins, 1997.

**Psychic Sciences**
Walter B Gibson and Litzka R Gibson
Souvenir Press, 1996.

**The History of Witchcraft & Demonology**
Montague Summers
Castle Books, 1992.

**Techniques of High Magic**
Frances King and Stephen Skinner
Sphere, 1997.

**Royal Fortune Teller**
Various authors
The Booksellers, 1995, previously 1945.

**Magical Book**
Nicole Sommesous
Thorsons, 1999.

**Born in Albion**
David Williams and Kate West
Pagan Media Ltd, 1996.

# index

# dedication

This spell is dedicated to myself, my family, my friends, all who were involved in producing this book and all who read it.  It is to the mighty Viking god Tyr.
The following prayer is taken from a booklet, Viking Magick Chants,
by Patricia Nelson, published by Finbarr International in 1988.
If you want to make a proper job of this, light one large, white candle
and repeat the following three times:

*'I call upon and summon you, O great and kind TYR,
to encircle me in your protective circle of light.
Be like a strong fortress for me, one that is impossible to break into;
so that no enemy can ever harm or touch me.
I will forever enjoy Thy protective embrace.
So be it!
Thank you oh mighty TYR, for listening to my plea.'*

# credits and acknowledgements

With grateful thanks to Jonathan Doo, Soldiy Bate, Nigel Dourne, Kule Rheeders, Barbara Ellen, Malcolm Wright and sundry long-departed relatives who passed on information about spells, charms, superstitions, history and a whole rag-bag of assorted information that has lodged itself inside my head for the past half-century.

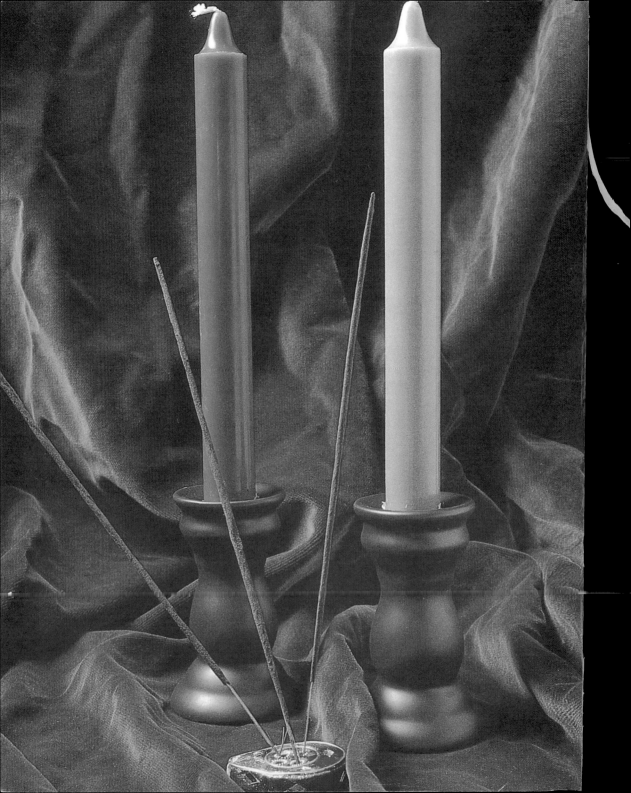

# the book of
# spells

D0531669